Postgrowth and Wellbeing

"Transition to a postgrowth economic order is now becoming much more widely accepted as essential to long-term social and environmental wellbeing. But the implications for the quality and provision of welfare have hitherto been little researched. This conceptually challenging but at the same time empirically grounded discussion provides a much-needed addition to the literature of the degrowth movement".

—**Kate Soper** (Emerita Professor of Philosophy at London Metropolitan University)

"Like it or not, growth may be coming to an end. How to sustain welfare without growth is a core question for our times. This book presents the boldest attempt to think what viable welfare systems a global steady state."

—**Giorgos Kallis** (Research Professor at the Autonomous University of Barcelona)

"At a time when leading economists are horrified by the spectre of secular stagnation affecting advanced capitalist economies, theorists of 'postgrowth' and a global 'steady-state economy' embrace the idea that its arrival might even enhance human wellbeing and needs satisfaction. But who knows what kind of views our future selves, even our descendants, may form about the requisites of wellbeing and the good life? Are we trapped in the social practices of mindless consumerism/productivism and GDP fetishism or ready to consider the worth of an economy that minimizes ecologically harmful levels of resource extraction and waste production? Büchs and Koch provide a concise, systematic and crystal clear account of what social sciences (including psychology and philosophy) have to offer on these questions—a critical account that is clinically clean of jargon and alarmist gesturing".

—**Claus Offe** (Professor Emeritus of Political Sociology at the Hertie School of Governance)

"Endless economic growth will destroy the planet, but a postgrowth economy threatens security and wellbeing. Büchs and Koch provide an honest and concise account of the dilemma and a guide to some ways forward."

—**Ian Gough** (Emeritus Professor of Social Policy at the University of Bath; Visiting Professor at the London School of Economics)

"Concise and well-written, this book offers a historical and comparative perspective on the changing relationship between wellbeing, economic growth and ecological perspectives as they intersect with the welfare state. Scholars, university students and activists interested in the relationship between capitalism, environmental policy and the welfare state should read this book."

—**Daniel Béland** (Canada Research Chair in Public Policy and Professor at the Johnson-Shoyama Graduate School of Public Policy)

Milena Büchs · Max Koch

Postgrowth and Wellbeing

Challenges to Sustainable Welfare

palgrave
macmillan

Milena Büchs
Sustainability Research Institute,
School of Earth and Environment
University of Leeds
Leeds, UK

Max Koch
Lund University
Lund, Sweden

ISBN 978-3-319-59902-1 ISBN 978-3-319-59903-8 (eBook)
DOI 10.1007/978-3-319-59903-8

Library of Congress Control Number: 2017944148

Printed on acid-free paper

This Palgrave Macmillan imprint is published by Springer Nature
The registered company is Springer International Publishing AG
The registered company address is: Gewerbestrasse 11, 6330 Cham, Switzerland

Acknowledgements

We would like to thank Judith Allan from Palgrave who one day in late 2015 unexpectedly dropped by Milena Büchs' office to ask about publication ideas. When Milena spontaneously formulated the idea for this book, Judith, even more unexpectedly, strongly supported it. Without her encouragement, this book would not exist. Many thanks also to Beth Farrow who took over from Judith in the process of writing the book and who provided fantastic support. In an effort to find a colleague to share the work, Milena asked Max Koch whether he wanted to come on board. Fortunately, he did not need much persuasion and contributed to the project in equal measure.

The seeds for the idea for this book were sown a few years back. For Milena, the discussions about relationships between energy, material flows and planetary wellbeing with colleagues from the EPSRC-funded project "*Transforming the Engineering of Cities to Deliver Societal and Planetary Wellbeing*" (Liveable Cities, EP/J017698/1) were especially important, including John Urry, Jane Falkingham, 'Bakr Bahaj, Patrick James, Luke Blunden, Leo Bourikas, Phil Wu, Chris Rogers, Helene Joffe and Chris Boyko, who also commented on a draft chapter. Many thanks also go to the German Research Foundation funded Postgrowth Societies College in Jena, which granted her a one-month fellowship in 2014. Milena's contribution to this book draws to a large extent on research conducted during this period and benefited from discussions with colleagues there, especially Stephan Lessenich and Dennis

Eversberg. Finally, Milena is very grateful for the support and comments provided by her colleagues at the Sustainability Research Institute at the University of Leeds, especially Daniel O'Neill, Julia Steinberger and Lucie Middlemiss. Milena would like to dedicate the book to Richard Douthwaite, who she met through her engagement with the Foundation for the Economics of Sustainability (Feasta) in the late 2000s. Richard passed away far too early in 2011. He, as well as other friends and colleagues from Feasta, especially Nick Bardsley and Brian Davey, have been a great inspiration for this work.

Max Koch wishes to acknowledge that his contribution to the book benefited from funding of FORTE (*Sustainable Welfare and Eco-social Policies*, 2016-07284), FORMAS (*The New Urban Challenge? Models of Sustainable Welfare in Swedish Metropolitan Cities*, 2016-00340) and the Research Council of Norway (*Sustainable European Welfare Societies: Assessing linkages between social and environmental policy*, 236930/H20). His special thanks go to the *Sustainable Welfare* and Degrowth research groups at Lund University and to the Social Policy subject area of the University of Edinburgh which hosted Max in the spring term of 2017. Much of his contribution was written during this visiting fellowship.

Contents

About the Authors

Milena Büchs is Associate Professor in Sustainability, Economics and Low Carbon Transitions at the University of Leeds, UK. Over the past decade, her research has focused on social dimensions of low carbon transitions, including distributional implications of climate change mitigation policies; the role of the voluntary sector in supporting low carbon practices; and the role of values for practice change.

Max Koch is Professor of Social Policy at Lund University (School of Social Work), Sweden. His research addresses capitalist restructuring and its implications for social inequality, welfare and employment relations. More recently, he has combined these research interests with issues of ecological sustainability, particularly climate change, sustainable welfare and postgrowth societies.

Abbreviations

ABC	American Broadcasting Company
AIDS	Acquired Immune Deficiency Syndrome
CBS	Columbia Broadcasting System
CO_2	Carbon dioxide
COP	Conference of Parties
EPSRC	Engineering and Physical Sciences Research Council
EROEI	Energy returned on energy invested
EU	European Union
FDI	Foreign Direct Investment
FORMAS	Forskningsrådet för miljö, areella näringar och samhällsbyggande (Swedish Research Council Formas)
FORTE	Forskningsrådet för hälsa, arbetsliv och välfärd (Swedish Research Council for Health, Life and Welfare)
GDP	Gross domestic product
GNP	Gross national product
GPI	Genuine Progress Indicator
HIV	Human Immunodeficiency Virus
IEWB	Index of Economic Well-Being
IPCC	Intergovernmental Panel on Climate Change
ISEW	Index of Sustainable Economic Welfare
NASA	National Aeronautics and Space Administration
NBC	National Broadcasting Company
NEF	New Economics Foundation
OECD	Organisation for Economic Cooperation and Development
SSE	Steady-state economy

UK	United Kingdom
UNCTAD	United Nations Conference on Trade and Development
UNFCCC	United Nations Framework Convention on Climate Change
US	United States of America

List of Figures

List of Tables

CHAPTER 1

Introduction

Abstract The era of economic growth is increasingly being questioned. Not only do economic systems in the developed world seem to be afflicted by "secular stagnation", but a departure from growth is also beginning to be presented as a moral imperative to safeguard human and planetary wellbeing in the long run. This has far-reaching implications for wellbeing which is currently coupled with a range of institutions, including the welfare state, organised around the growth paradigm. This book critically discusses the assumption in the postgrowth literature that wellbeing can be maintained or even improved without growth. It highlights ways in which theories of human need, social practices and political economy can contribute to this debate.

Keywords Postgrowth · Wellbeing · Social practices · Political economy

The era of economic growth is increasingly being questioned. This is for two reasons. First, the *capacity* for growth seems to diminish amongst most Western economies as growth rates have been persistently falling since the 1980s. Leading economists have proclaimed a new era of "secular stagnation" (e.g. Summers 2016) for which they have identified a variety of possible reasons, reaching from lacking investment in infrastructure and education, a related savings glut, lack of innovation and decline of productivity, demographic changes (ageing societies due to rising life expectancy and lower birth rates), high government and private

M. Büchs and M. Koch, *Postgrowth and Wellbeing*,
DOI 10.1007/978-3-319-59903-8_1

debt, and increasing inequality (Summers 2016; Gordon 2012; Streeck 2014). Second, the *justifications* for growth as the primary end of economic activity have become discredited because of its many negative consequences for environment and society. It is increasingly being recognised that economic growth will threaten human and planetary wellbeing in the long term as we have already overstepped several ecological thresholds (Rockström et al. 2009) and are depleting vital resources. In many ways, we may therefore already be in the process of entering a new era of postgrowth.

These insights have far-reaching implications because our current societies and welfare systems are built upon growth. While a departure from growth increasingly looks like an inevitable development inherent in current economic systems, as well as a moral imperative to safeguard human and planetary wellbeing in the long run, it is of vital importance to think through potential repercussions for human wellbeing in the short and long term. To what extent does human wellbeing depend on economic growth—and growth-based welfare provision—within current economic and social systems? Which insights can we draw from existing evidence and social theory on the ways in which economic growth and contraction, and social change more generally, influence wellbeing? What kinds of conditions are likely to be required to secure human wellbeing under postgrowth in the long term, and which institutions might be able to support them? These are some of the main questions that this book is concerned with.

The literature on postgrowth is rapidly expanding. By postgrowth, we refer to two interlinked ideas: the concept of a steady-state economy (SEE) (Daly 1972; Daly and Farley 2011) and the discourse around degrowth, which can be seen as a transitionary phase of reaching an SSE (Kerschner 2010). The maintenance or even improvement of human wellbeing has been defined as one of the core aims of postgrowth (Schneider et al. 2010), and postgrowth supporters seem optimistic that this aim can be achieved. This may be a necessary standpoint to take from the strategic point of view of a social movement which is currently seeking wider public support. However, for it to become a realistic and successful idea, we believe a more critical and empirically informed discussion of the relationships between postgrowth and wellbeing is required.

The literature on wellbeing commonly draws a distinction between subjective and objective dimensions of wellbeing. Arguments that support the view that wellbeing can be maintained or improved in the context of postgrowth have been made in relation to both dimensions. First of all, the postgrowth literature criticises the equivalisation of wellbeing with GDP. Instead, it encourages a focus on broader notions of wellbeing which relate to ideas such as flourishing, the good life or alternative notions of prosperity (Schneider et al. 2010: 513; Muraca 2012; Jackson 2011; Kallis 2011: 879), all of which put less emphasis on material living standards. Furthermore, postgrowth supporters point to evidence that beyond a certain income threshold, subjective and objective wellbeing no longer increase, or even decrease, with additional income gains—where subjective wellbeing is usually measured through happiness or life satisfaction surveys and objective wellbeing with alternative indicators such as the Genuine Progress Indicator or the Human Development Index. Postgrowth advocates argue that this evidence indicates that reductions of GDP are unlikely to decrease wellbeing, or could even improve it, because zero growth or contraction could provide more time and opportunities to support the things that people need to flourish—supportive relationships, time for leisure and meaningful work, community involvement, etc.

These are very valid arguments. However, more discussion is required to see how confident we can really be based on this evidence that wellbeing can be maintained or improved in the context of postgrowth. This applies especially to the phase of degrowth which will involve a contraction of the economy and a decrease of material living standards in rich Western countries. Some research has shown that while people do not strongly respond emotionally to welfare gains, they do respond—negatively—to welfare losses (Tversky and Kahneman 1991). In addition, there is also evidence at the collective (national) level that people's subjective (Fanning 2016: 100; Easterlin 2009) and objective wellbeing (here mainly measured in terms of health outcomes) (Gavrilova et al. 2000; Notzon et al. 1998; Men et al. 2003) declines in phases of economic contraction. Certain wellbeing outcomes such as health and education are likely to be negatively affected by austerity measures introduced during times of economic contraction. Postgrowth proponents will argue that economic crises in the current system differ from voluntarily induced, planned phases of degrowth or new systems of SSEs which would feature new institutions to prevent wellbeing losses. This is an important objection which also points to one of the main difficulties that this book is confronted with—the fact that the institutional set-up of future socio-economic systems—and their interaction with wellbeing outcomes—cannot be foreseen. However, wellbeing implications of declining or stagnating economic growth can be studied empirically, and it remains important to discuss which insights this provides regarding the generation of wellbeing in the context of postgrowth.

A first argument that we will advocate in the book is that further discussion is required about the conceptions of wellbeing that should be applied in these debates. At present, the discussion of wellbeing in the postgrowth literature remains incoherent as different scholars refer to different concepts and measures, also depending on the type of argument they want to make. Some refer to narrow concepts of happiness or life satisfaction, others to broader notions related to eudemonic wellbeing. Others yet have focused more on alternative welfare indicators, taking an objective perspective. We argue that notions of basic human needs deserve greater emphasis in debates on postgrowth because they are compatible with ideas about limits to growth. While public debates have routinely made connections between the notions of human needs and concern for the wellbeing of future generations (inspired by the Brundtland report (WCED 1987)), we argue that the conceptual links between these ideas still need to be worked out more systematically. Moreover, we argue that the objective approach to wellbeing represented by the basic human needs approach should acknowledge more explicitly the relevance of subjective dimensions of wellbeing because these two dimensions are interdependent.

Second, we argue that we do not yet understand very well, conceptually and empirically, the multiple ways in which rapid socio-economic change, in particular degrowth/economic contraction as a specific case of such change, affects subjective and objective wellbeing. Sociologically, we can conceptualise the generation of wellbeing as an outcome of social practices which mediate between agency and structure dimensions of society and hence continuously reproduce, stabilise or change society. From this perspective, socio-economic transitions will inevitably require changing structures at both societal and actor levels (e.g. distribution of resources, institutions, discourses at the social level; and worldviews, values, competences and other dispositions at the actor level), as well as technologies/infrastructures. It is likely that changes in these multiple dimensions, which are all in some way or other tied up with the current economic system, need to be well aligned with each other to enable the maintenance or improvement of wellbeing in phases of degrowth and rapid societal transition more generally. Therefore, problems may occur if these changes do not operate at similar time scales or if they are in other ways out of sync. An example is the need for rapid cultural changes alongside economic change: people's values and conceptions of wellbeing would need to transform in the move away from a growth-based consumer society to avoid perceptions of deteriorating living standards, sacrifice, reversal of social progress, etc., and the negative wellbeing implications that such perceptions might promote.

The purpose of discussing these issues is not to argue against postgrowth. We wholly accept this position as our starting point but stress that we need to better understand the ways in which wellbeing may currently be intertwined with economic growth and (welfare state) institutions which support growth, all of which are closely coupled to market capitalism. We hope this will provide insights into required social and institutional changes to decouple wellbeing and growth. For instance, there is some evidence that greater social equality contributes positively to several dimensions of wellbeing, including health, social capital and trust (Wilkinson and Pickett 2009), and it is plausible that it could minimise possible negative effects of economic contraction on people's wellbeing. This is one of the reasons why several authors have emphasised that postgrowth would need to be framed by more redistributive institutions to maintain or even improve wellbeing (e.g. Schneider et al. 2010).

One of the hotly debated questions in this context is whether the institutional changes that are required to support the transition to a low-carbon postgrowth economy can be achieved within current systems of global market capitalism or whether all of the necessary changes taken together would generate a qualitatively different system. This question cannot be answered categorically, and to some extent, it will depend on the definition of market capitalism that is applied. However, since it is crucial to understand the various ways in which the generation of wellbeing is embedded in existing socio-economic systems, Chap. 2 of this book traces the relationship between the emergence of the current growth paradigm and the development of capitalism. It will show that economic growth is a fairly recent phenomenon which is inherently linked to the emergence of capitalism in the seventeenth and eighteenth centuries. The chapter also discusses some of the sources of the contemporary crisis of market capitalism and how these may undermine the prospects for future growth.

Chapter 3 turns to the discussion of the evolution of welfare states in Western capitalism—one of the (bundles of) institutions that have become so relevant for supporting many aspects of people's wellbeing. This chapter demonstrates the close links between market capitalism, economic growth and welfare state development which has important repercussions for debates about postgrowth. It also asks whether there is a relationship between welfare state development and environmental protection—which will be fundamental for wellbeing in the long term. However, it concludes that environmental destruction remains primarily driven by GDP growth and that welfare institutions have done little to change this relationship.

Chapter 4 provides a brief summary of arguments against growth. It shows that these debates go far beyond a criticism of GDP as a measure of welfare by highlighting the problematic ecological and social consequences of growth. It also argues that the more recent discussion about limits to growth has moved away from a focus on resource, especially fossil fuel, limitations. Instead, concerns about overshooting a range of "planetary boundaries" (Rockström et al. 2009), especially potentially catastrophic and irreversible effects of climate change, have become more important for thinking about limits to growth, as it is already affecting people's wellbeing and will fundamentally undermine it in the long run if left unchecked. In the final part of this chapter, we introduce ideas of postgrowth which have emerged from growth-critical debates. Here, we provide an overview of different positions within this field by distinguishing system-reform, anti-capitalist and alternative-open approaches which differ in relation to the ways in which they criticise growth and consider the relationship between capitalism and growth, and the visions they put forward for postgrowth societies.

Chapter 5 reviews in more depth the discussion about the relationship between postgrowth and wellbeing. After providing a brief overview of different wellbeing concepts and measurements, it presents and critically discusses the main arguments that have been made in the postgrowth literature regarding the capacity of postgrowth economies to maintain or even improve present levels of wellbeing. This involves a discussion about the concepts of wellbeing that have been applied in the debate so far, as well as a review of the evidence on the relationships between economic growth and contraction on the one hand and subjective and objective wellbeing outcomes on the other. The chapter argues that the concept of basic human needs deserves more attention in this debate as it is compatible with postgrowth frameworks.

Chapter 6 proposes that social practices theory is a useful lens for conceptualising possible wellbeing implications of postgrowth. Highlighting the interdependency of structure and agency in the generation of wellbeing, this chapter reflects on the ways in which different social dimensions—e.g. resources, institutions and discourses—interact in generating wellbeing at individual and social levels and the roles that stability and change play within these processes. This is relevant because at a more general level, the transition to sustainable postgrowth will need to involve very profound and rapid social changes, often of various social structures in parallel. An important question emerging from this is whether the various social structures that are currently organised around market capitalism and its inherent power structures can change rapidly enough and at similar speeds to avoid deteriorations of wellbeing, including the important role of cultural change.

Chapter 7 takes the debate back to a more concrete level and asks which kinds of institutions could support wellbeing in the context of sustainable postgrowth, especially in periods of far-reaching social change. We start with a discussion of core principles for the achievement of an SEE that are being discussed in ecological economics—including a sustainable scale of material throughput and social equality, followed by considerations of the role of spatial scales—the requirement to share and coordinate responsibility for transitions towards postgrowth across global, national, regional and local levels, as well as of a range of more fine-grained policy proposals that have been made to support wellbeing under postgrowth, focusing on macroeconomic steering, inequality/redistribution, minimum and maximum incomes, carbon rationing, consumption, working time reduction, work–life balance as well as population/migration.

References

Daly, H. 1972. In Defense of a Steady-State Economy. *American Journal of Agricultural Economy* 54 (5): 945–954.

Daly, H., and J. Farley. 2011. *Ecological Economics. Principles and Applications*, 3rd ed. Washington: Island Press.

Easterlin, R.A. 2009. Lost in Transition: Life Satisfaction on the Road to Capitalism. *Journal of Economic Behavior & Organization* 71 (2): 130–145.

Fanning, A.L. 2016. Policy Options for Sustainable and Equitable Coastal Economies: A Comparative Case Study. Doctoral Thesis, University of Cadiz.

Gavrilova, N.S., V.G. Semyonova, G.N. Evdokushkina, and L.A. Gavrilov. 2000. The Response of Violent Mortality to Economic Crisis in Russia. *Population Research and Policy Review* 19 (5): 397–419.

Gordon, R. J. 2012. Is US Economic Growth Over? Faltering Innovation Confronts the Six Headwinds. Working Paper 18315. Cambridge: National Bureau of Economic Research.

Jackson, T. 2011. *Prosperity without Growth: Economics for a Finite Planet.* London: Earthscan/Routledge.

Kallis, G. 2011. In Defence of Degrowth. *Ecological Economics* 70: 873–880.

Kerschner, C. 2010. Economic De-Growth vs. Steady-State Economy. *Journal of Cleaner Production* 18 (6): 544–551.

Men, T., P. Brennan, P. Boffetta, and D. Zaridze. 2003. Russian Mortality Trends for 1991–2001: Analysis by Cause and Region. *British Medical Journal* 327 (7421): 964–966.

Muraca, B. 2012. Towards a Fair Degrowth-Society: Justice and the Right to a "Good Life" Beyond Growth. *Futures* 44 (6): 535–545.

Notzon, F.C., Y.M. Komarov, S.P. Ermakov, C.T. Sempos, J.S. Marks, and E.V. Sempos. 1998. Causes of Declining Life Expectancy in Russia. *Journal of the American Medical Association* 279 (10): 793–800.

Rockström, J., W. Steffen, K. Noone, A. Persson, F.S. Chapin, E.F. Lambin, T.M. Lenton, M. Scheffer, C. Folke, H.J. Schellnhuber, B. Nykvist, C.A. de Wit, T. Hughes, S. van der Leeuw, H. Rodhe, S. Sorlin, P.K. Snyder, R. Costanza, U. Svedin, M. Falkenmark, L. Karlberg, R.W. Corell, V.J. Fabry, J. Hansen, B. Walker, D. Liverman, K. Richardson, P. Crutzen, and J.A. Foley. 2009. A Safe Operating Space for Humanity. *Nature* 461 (7263): 472–475.

Schneider, F., G. Kallis, and J. Martinez-Alier. 2010. Crisis or Opportunity? Economic Degrowth for Social Equity and Ecological Sustainability. Introduction to this Special Issue. *Journal of Cleaner Production* 18 (6): 511–518.

Streeck, W. 2014. How Will Capitalism End? *New Left Review* 87: 35–64.

Summers, L.H. 2016. The Age of Secular Stagnation What It Is and What to Do About It. *Foreign Affairs* 95 (2): 2–9.

Tversky, A., and D. Kahneman. 1991. Loss Aversion in Riskless Choice: A Reference-Dependent Model. *The Quarterly Journal of Economics* 106 (4): 1039–1061.

WCED. 1987. *Our Common Future (Brundtland Report)*. Oxford: Oxford University Press for World Commission on Environment and Development.

Wilkinson, R.G., and K.E. Pickett. 2009. *The Spirit Level. Why More Equal Societies Almost Always Do Better*. London: Allen Lane.

CHAPTER 2

Capitalist Development and the Growth Paradigm

Abstract This chapter traces the relationship between the emergence of the current growth paradigm and the development of capitalism. It argues that economic growth is a fairly recent phenomenon which is inherently linked to the emergence of capitalism in the seventeenth and eighteenth centuries. It argues that economic growth in capitalism is inevitable, since this economic system is oriented towards unlimited and short-term valorisation, quantitative and geographic expansion, circularity and reversibility. This monetary or "exchange value" aspect of the capitalist economy is in later chapters contrasted with the principles that guide the ecological system (the "use-value" aspect), involving stable and sustainable matter and energy transformations and throughputs as well as irreversibility.

Keywords Capitalism · Growth · Profit motive · Stagnation

According to environmental historian McNeill (2000: 236), the "overarching priority of economic growth was easily the most important idea of the twentieth century". And today, in the early twenty-first century, what Herman Daly (1972) first called the "growth paradigm" is almost universally accepted. This paradigm presupposes that economic growth is "good, imperative, essentially limitless, and the principal remedy for a litany of social problems" (Dale 2012a). The predominant approach in economics, the neoclassical perspective, tends to identify prosperity

M. Büchs and M. Koch, *Postgrowth and Wellbeing*,
DOI 10.1007/978-3-319-59903-8_2

with not merely wealth but growing wealth (Soper and Emmelin 2016). It views economics as a repetitive cycle linking money and commodities as well as households and companies. A "return to capital" basically means that the original capital spent, augmented by a surplus, returns to its owner, and the process of capital valorisation starts over again on a greater scale. The production of goods and services is analysed from the standpoint of growth of monetary value, which is seen as indefinite, while the roles played by energy and natural resources in this production are sidelined. Hence, the economy is conceptualised as if it were a closed system, within which flows of services and goods are compensated by financial flows in the opposite direction and whose coherence is guaranteed by the link of exchange alone, while use values, matter, energy and nature in general are treated as if they were infinite. However, economics has not always been regarded as synonymous with a science of prices, exchange value and monetary growth. In this chapter, we trace the development of the "growth paradigm" by focusing on its parallel development with capitalism. How did the notion of economic growth emerge and under what socio-economic conditions did it become hegemonic? What are the prospects for future growth?

Economic Growth in the Pre-capitalist World

Angus Maddison (2007) empirically demonstrates that before the 1820s, when economic growth started to accelerate in the context of the industrial revolution, global economic activity had been characterised by periodic swings, but expanded by an average of 0.05% annually only, and this was largely due to a slow increase in populations. Ancient civilisations knew commitments to the accumulation of wealth, especially the expansion of territory and riches earmarked for particular purposes such as the building of palaces or pyramids. The "impulse to acquisition, pursuit of gain, of money, of the greatest possible amount of money" has, as Max Weber (1958: xxxi–xxxii) famously pointed out, "in itself nothing to do with capitalism" and "exists and has existed among waiters, physicians, coachmen, artists, prostitutes, dishonest officials, soldiers, nobles, crusaders, gamblers and beggars". While, hence, unlimited greed for gain is "not in the least identical with capitalism", the "pursuit of profit", and particularly that of "forever renewed profit, by means of continuous, rational, capitalist enterprise" (ibid.) indeed is. In feudal societies, by contrast, the pursuit of profit for its own sake tended to be seen as deviating from the norm. In medieval Europe, for example, economic interests tended to be subordinate to what Weber (1958) referred to as "salvation" (Fig. 2.1).

In the sixteenth- and seventeenth-century England, the societal respectability for the pursuit of "forever renewed profit" (Weber) grew, and this was reflected in increasingly liberal trade regulations. However, the concern of mercantilists was "not growth in production for use but the increase in products for sale" (Dale 2012a) with the expansion of exports becoming a "state-supported imperative". Not growth per se was the goal but the "enrichment of the state. ... Acquisition was what mattered, not production or consumption" (Dale 2012a). In the pre-capitalist world, most economic activity—agrarian labour—and time followed daily and seasonal solar cycles. Mark Elchardus (2011: 15) recalls that before the introductions of the Gregorian calendar by Pope Gregory XIII, which corrected some deficiencies of the Julian calendar and standardised time in the Christian world, and, specifically, the Greenwich

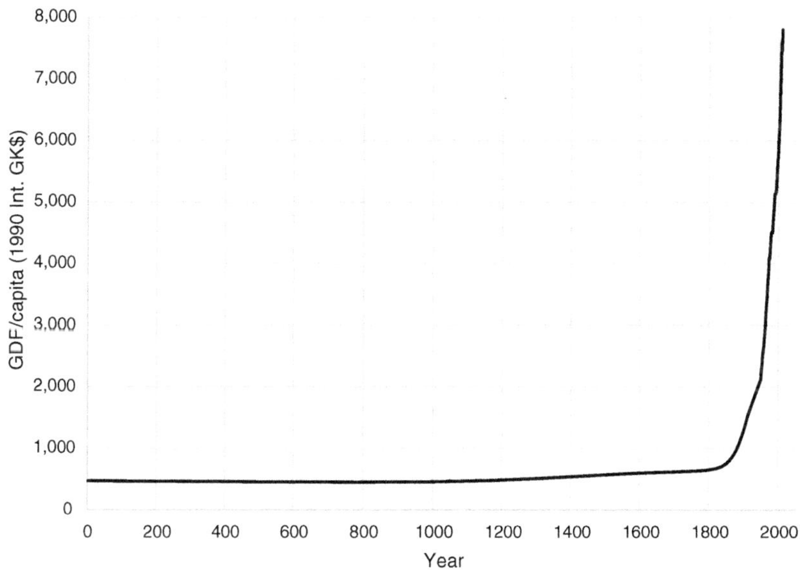

Fig. 2.1 World GDP per capita 1–2010 (1990 Int. GK$). *Source* The Maddison-Project Historical Database, http://www.ggdc.net/maddison/maddison-project/home.htm (2013 version)

Mean Time, most localities had a local time or a mean local time, based on the sun. This meant "that in general every difference of 20 km in the east–west direction corresponded to a time difference of about 3 minutes". The standardisation of time went hand in hand with the spatial unification of capitalist markets and, specifically, the development of transport, particularly railroads and shipping. In 1875 representatives of the railroads and shipping interests proposed to unify the world in one time reckoning system: "using one reference point and creating zones that differed in full hours from that reference point" (Elchardus 2011: 15). Once the Greenwich Mean Time had been adopted as the international standard in 1885, time could "appear as an abstract continuum, uniform, linear and measurable" (Dale 2012a).

Political economists of the pre-industrial period did not conceive growth in abstract, quantifiable terms, or as a principle policy goal for governments. The economy was instead presented as processes that follow a natural rhythm. In the Physiocratic system, the wealth of nations was derived solely from the value of land and the entire economic process was understood through focusing on the productivity of agriculture that was seen as the only kind of work that created value and surplus (Cleveland 1999: 127). In his *Wealth of Nations*, Adam Smith (1993) was the first political economist to suggest a direct link between a nation's welfare and the amounts of goods and services produced. To support this, he encouraged free trade that would advance the division of labour which, in turn, would lead to further specialisation. While, hence, "Smith did more than anyone to elaborate a conception of economic growth as natural, self-reinforcing, and an unqualified good" (Dale 2012a), this advocacy of self-sustaining growth was somewhat tempered due to his consideration of countervailing tendencies such as heightened competition amongst large companies that would result in declining profit rates. David Ricardo was first to conceptualise the economy as a separate sphere with respect to politics. Like Smith, he assumed a long-term tendency of diminishing returns and growth rates due to an increasingly competitive context, but, unlike Smith, Ricardo argued that this could be postponed to "the 'almost indefinite future'" (cited in Dale 2015) by technical progress, foreign trade and the exploitation of overseas' resources.

In his *Principles of Political Economy*, and particularly with the notion of a "stationary state", John Stuart Mill likewise grapples with the issue of diminishing returns. As growth rates decrease, a "stationary economy"—one that does not grow further in monetary terms—would be the inevitable result: "It must always have been seen, more or less distinctly, by political economists, that the increase of wealth is not boundless: that at the end of what they term the progressive state lies the stationary state" (Mill 1848: 514). Mill was bold in predicting, in the mid-nineteenth century, that the "richest and most prosperous countries would very soon attain the stationary state, if no further improvements were made in the productive arts, and if there was a suspension of the overflow of capital into the uncultivated or ill-cultivated regions of the earth ..." (1848: 514). In contrast to Smith, who considered a stationary state a "dull" affair (cited in Dale 2012b: 865), Mill thought that falling profit rates would have a positive effect. Distinguishing between a stationary state of the economy and a "stationary state of human improvement", he argued that there would be "as much scope as ever for all kinds of mental culture, and moral and social progress; as much room for improving the Art of Living, and much more likelihood of its being improved, when minds ceased to be engrossed by the art of getting on" (Mill 1848: 515).

Gareth Dale's recent reconstruction of classical political economists and their respective contributions towards understanding environmental issues in the context of capitalist development and economic growth is of great merit given that Mill's "stationary state" became a common point of reference in present ecological economics, particularly in Daly (1972; 2011), Jackson (2009) and a range of degrowth texts (Chap. 4). Though Mill goes some way in explaining the growth imperative in capitalist society, Dale (2012b: 437) demonstrates that Mill gives "unqualified support to its basic institutions—wage labour, market exchange, and private property in the means of production—as well as to *laissez faire* and free trade". There is indeed a tension between Mill's support of a stationary state, on the one hand, and of an unfettered capitalism, on the other. The failure to link the two largely contributes to the "determinism" (Dale) from which Mill's projected falling rate of profits suffers. He therefore ends up not systematically considering what Marx called "countervailing tendencies" for the profit rate to fall, chiefly the intermediating and growth-promoting effects of technical and spatial fixes as well as the expansion of foreign trade. And instead of exploring "the participatory-democratic possibilities that a stationary state might afford" (Dale 2012b: 438), Mill promoted a number of authoritarian measures to enforce population control including the separation of spouses in work houses and the support for legislation forbidding marriage unless the parties could provide the means for supporting a household.

Growth as Capital Accumulation: Karl Marx

By contrast to his classical political economy predecessors, Karl Marx witnessed a more developed capitalism, where the industrial revolution had dramatically raised labour productivity and most labour products had taken the form of commodities, that is, they were produced for exchange on markets. Marx develops the notion of the structural imperative of capitalist economies to expand in scale and grow in monetary terms from the logic of exchange relations and, specifically, the money form (Marx 1961: 94–142). He compares two kinds of exchange or "metamorphoses" of commodity and money. In the first one—commodity–money–commodity—the purpose of the exchange is qualitative. A holder of a commodity exchanges it for its money equivalent; then, he or she buys another commodity for his or her own use: "selling in order to buy" (Marx 1961: 147). In this exchange, the role of money is that of a measure and store of value as well as that of a legal tender. Then, Marx (1961: 130) argues that with the "very earliest development of the circulation of commodities, there is also developed the necessity, and the passionate desire, to hold fast the product of the first metamorphosis": money serves here as the general and ultimate expression of the wealth available in a society or as capital that potentially leads to profit and bears interest. The purpose of the second metamorphosis—money–commodity–money—can only be a quantitative one, since there is no qualitative difference between its origin and result: the production of more money compared to the original amount.

According to Marx, profits can be made due to the fact that a commodity is available for sale that has the use value of creating exchange value and can be used longer than the time period that represents the cost of its own reproduction: labour power. In the capitalist mode of production, producers—as wage-earners—are largely separated from their means of subsistence and production and have no alternative but to offer the only commodity at their disposal on "labour markets". Likewise, the other "factors of production"—land, raw materials, fuels, auxiliaries, etc.—can be purchased on separate markets as "fictitious commodities" (Polanyi 1944), and it is only through the intermediation of employers, who hold the necessary capital, that the various elements of the production process come together. For capitalist production, all that matters is that these factors and the ingredients of material production are separately available for purchase and in forms that can be combined in the production process of capital.[1]

Marx (1961: 312–321) discusses the tendency of capitalist economies to expand in scale—and thereby the "growth imperative"—when analysing the production of "relative surplus value". The profitability of a company can not only be improved by increasing the working hours of the wage-earners ("absolute surplus value") but also by shortening the part of their working day that is necessary for the workers' physical and social reproduction. Marx explains a reduction in the price of labour power through increases in productivity in those branches of production that are part of the consumption patterns of the wage-earners. However, he also stresses that the realisation of such a relative surplus value will face an immanent contradiction: individual owners of capital are permanently motivated to optimise the technological and organisational basis of the work process in order to be one step ahead of their competitors. This is normally carried out by a substitution of workers by machinery or by an improved organisation of the internal division of labour. The employers whose productivity level is above average can thus achieve extra profit since they are able to sell their commodities at prices below the normal level.

Yet such an improvement of production methods tends towards generalisation, and the extra profit moves towards zero, since competing companies have no choice but to copy the new methods or even to improve upon them. As the new productivity level gradually becomes the new social standard, a given quantity of commodities is now produced with less labour effort than previously—and the price of a single commodity decreases as a result. Marx concludes that, on the one hand, the rate of surplus of the employed workers increases (because of higher volumes of sales per worker); on the other hand, however, the absolute volume or mass of surplus value (and, other conditions being equal, the mass of profit) decreases since fewer workers are needed to produce a given amount of commodities than before. In order to keep the volume of profit stable, despite this dilemma, there is no alternative but to expand the overall scale of production through the reinvestment of previous profit, in other words "accumulation" of capital.

Marx was aware of the structural tensions that exist in an economy geared towards growth of money as a homogenous material entity and the general principles of the work process based on heterogeneity of its natural and material ingredients the combination of which is bound up with rearrangements of energy and matter (Burkett 1999; Koch 2012). Using the historical example of England, he discussed the advancement of the division of labour and how the work process became successively independent from the individual skills of workers through the systematic application of natural forces and the natural sciences. The Industrial Revolution introduced tools and machinery that reduced the role of many individual workers to that of an "appendage". Once the work process had an industrial foundation, the subjugation of nature under capital became more complete. Expanding scales of production and economic growth coincided with greater amounts of throughput of raw materials and auxiliary substances, especially in the form of fossil fuels as well as of available energy. Rising demand for raw materials and available energy normally leads to rising prices, for example, for crude oil, creating incentives for individual companies to recycle and to use a given quantity of materials or fuels in more efficient ways (Marx 2006, Chap. 5).

Yet progress in the efficiency of raw and auxiliary materials does not fundamentally alter the link between the expansion of the scale of production and the increase in the material and energy throughput, a phenomenon that had been observed by William Stanley Jevons (1865). According to the "Jevons paradox", greater efficiency in the use of a fossil energy source such as coal or oil leads to an increase in demand—not to a decrease—and in fact constitutes a necessary precondition for further capital expansion and economic growth (see also Chap. 4). The second and third volumes of *Capital* reflect the fact that capital does exist not only in its productive, that is, value-producing, form but also in unproductive forms, as money and commodity capital. While alternating between these three forms, competition forces individual companies to reduce the two unproductive functions of the capital cycle and, hence, to speed up the overall turnover process as much as possible. Hence, while the matter and energy transformation processes associated with all work processes have an irreversible and linear character, the structural imperative towards increasingly rapid turnover cycles is characterised by "time–space compression" (Harvey 1990) and a tendency towards temporal and geographic "simultaneity".

Marx's *Critique of Political Economy* not only provides a theory of the structural imperative of capitalist economies to expand in scale and grow in monetary terms, it also demonstrates how the economic categories and social relations are reflected in actors' minds. The historically specific mode of transfer of surplus labour and its social genesis—specifically the fact that the appropriation of past unpaid labour is the prerequisite for the appropriation of further unpaid labour on an increasing scale—is hidden by a range of mystifications as a result of which the capitalist mode of production appears as the natural and eternal way of organising "the" economy. Due to the wage form, all labour seems to be paid so that profit in its various forms seems to result from other sources than surplus labour. The "topsy-turvy world" of the "trinitary form" (Marx 2006), where wage labour contributes to societal wealth on the same footing and in functional harmony with profits and rent is the structural context for the widespread idea that economic growth is beneficial to all—including to those who contribute to it through work. The corollary is the meritocratic illusion that the more one works, the greater one's share in societal wealth will be. In fact, from the common perspective of commodity–money transactions, own work seems to be the only possibility of becoming a commodity owner in the first place. Core societal values and orientations such as "achievement", "upward mobility" and "social position as result of own work and merits", which are of crucial significance for the maintenance of the growth paradigm, have their structural basis in the specifically historical features of the capitalist production and accumulation process that present themselves as natural features of economic activity.

The Twentieth and Early Twenty-First Centuries

The OECD proclaims that for "a good portion of the twentieth century there was an implicit assumption that economic growth was synonymous with progress: an assumption that a growing Gross Domestic Product (GDP) meant life must be getting better" (OECD 2008, cover text). In the course of the nineteenth century, the reduction of concrete use values, matter and energy to abstract numbers and monetary magnitudes had begun to become a salient feature of economic life. In the early twentieth century, this development reached a new level, when, in 1932, the US Congress commissioned the economist Simon Kuznets to devise a means by which to measure the nation's output. This resulted in Gross National Product (GNP), a measure that estimated the market value of all final goods and services produced within a country per year, including the costs of government services. After the Second World War, GNP was turned into an official measure of economic policy in the USA. In 1953, the United Nations issued its international standards for a system of national accounts. In 1991, after the collapse of the Soviet Union which had used "net material product", which included physical goods but excluded services, as measure of economic progress, GNP was replaced by GDP. "With GNP, the earnings of a multinational company are attributed to the country where the company is owned, and where the profits end up. With GDP, on the other hand, the profits are attributed to the country where the factory is located and resource extraction occurs, even if the profits leave the country" (O'Neill 2013: 104).

Focusing on monetary flows, GDP does not distinguish between "good" and "bad" economic activities. While the purchases of beer, bicycles and cars as well as government investments in education contribute to GDP, it excludes various social practices relevant to human well-being including voluntary work and unpaid housework but also illegal transactions or environmental damages (see, for more details, Chap. 4). Despite these anomalies, GDP came to be seen as a proxy for the profitability of national economies and a sort of magic potion to cure all kinds of social problems during the so-called golden age of capitalism (1950–1975). For the USA, Dale (2012a) reports that in 1958 Henry Kissinger chaired a panel of economists representing large corporations and major universities. It produced a book called *The Key Importance of Growth to Achieve National Goals*, which identified "growth as the solution to the continual pressure of competing claims on national income (the arms race, public infrastructure, education, etc.)". Not only would economic growth bring "'dignity, freedom, and purpose'" (cited in Dale 2012a), but also expand the opportunities in combating poverty, economic hardship and poor health and in improving the educational system.

The regulation approach refers to the predominant growth strategy of this "golden" period as "Fordism" (Boyer and Saillard 2002; Koch 2006, 2012), a label that alludes to the division of labour in Henry Ford's automobile factory first used by Antonio Gramsci (1971). This growth model was characterised by a parallel restructuring of both the technological and organisational basis of the production process and the lifestyles and consumption patterns of wage-earners. It took the form of a compromise or exchange between management and organised labour: Wage-earners could benefit from productivity gains via wage increases, shortening of labour hours and the establishment and expansion of welfare services (Chap. 3). In return, trade unions accepted "scientific", that is Taylorist, management methods involving a clear distinction between conception and execution, production and sales, marketing and finance, where manual workers' function was largely reduced to simple and repetitive tasks within the work process, while skills, control assets and qualifications were increasingly concentrated within the planning department.

Originating in the USA, where the wartime experience of ending the depression seemed to justify the continuation of the "grand coalition" of government and business to prevent the recurrence of crises and to maintain economic growth (Agnew 1987), the new growth strategy began to be applied in Western Europe with the help of special economic recovery programmes such as the Marshall Plan launched in 1949. One favourable condition of the post-war period for economic growth to pick up was the fact that producers could count on a quasi-"infinite" demand for mass-produced goods such as automobiles and household appliances such as televisions and washing machines. Unlike the 1930s, when solvent consumers were scarce, during the era of post-war reconstruction, there was stable and expanding demand for both consumer goods and the means of production to build them. Since most Western European households did not yet own durable goods such as household appliances, mass production could become the technological basis for their speedy generalisation. The turnover of fixed capital was accelerated by the continuing increase in the number of products, which reduced the costs of one single product. Profits were supported by consumer demand, based on increasing real wages, which were usually determined by collective agreements and tied to expected growth in productivity. The result was unprecedentedly fast growth rates of GDP and productivity during the 1950–1973 period (Koch 2013: 33).

In the course of the 1970s, the Fordist growth strategy went into crisis.[2] Compared to the 1950s and 1960s, labour productivity and GDP growth fell in all countries on both sides of the Atlantic from levels of between 4 and 5% to around 1% in the 2000s (Koch 2013: 33). To understand how the crisis of Fordism was overcome and how a new period of capitalist growth was initiated, a number of scholars focus on the notion and process of the "financialisation" of socio-economic relations (Boyer 2000; Stockhammer 2008; Krugman 2009; Stiglitz 2010; Koch 2012). The term covers a range of phenomena including the deregulation of the financial sector and the liberalisation of international capital flows, with a corresponding increased instability in currency markets. It further reflects significant increases in financial transactions and the proliferation and profitability of new financial instruments such as hedge funds. The priority within companies' competitive strategies moves from investments in the real economy towards greater importance of financial profits, financial markets and foreign investment. In relation to consumption, wages and salaries continue to be essential for the demand generated by wage-earners, but they are increasingly complemented by loans (Boyer 2000), especially via mortgage-based borrowing. The decrease in real wages and the corresponding fall in spending power of wage-earners are partly compensated by the increase in the access of the wage-earning class to consumer loans. Money is issued primarily as debt, and it is now commonly held that, even more so than under Fordism, the economy as a whole needs to continue to grow so that debtors can service the growing volume of interest on the debt. Productive investment expenditures tend to be slow due to shareholder-value orientation and the general focus on financial profit. Such investment becomes more risky and is carried out under the imperative of sustaining higher profit rates than those achievable by financial investment. Generally, the hierarchy of institutional forms changes from a "management–labour balance" to a "management–shareholder balance" (Stockhammer 2008: 191). Last but not least, the cancellation of the Fordist compromise and its replacement by a finance-driven regime is not viewed as an exogenous shock to economies by most political economists but as the "outcome of particular policy arrangements" (Stockhammer 2008: 187). The overwhelming majority of countries introduced reforms that facilitated rather than complicated foreign and particularly financial investment since 1992 (UNCTAD 2009; Koch 2012: 97).

Under these new regulatory and institutional conditions, the continued search for growth is confronted with a range of "headwinds" (Gordon 2012) that include the interplay of globalisation and modern technology, which "accelerates the process of catching up of the emerging markets and the downward pressures on wages and real incomes in the advanced nations" (ibid.: 20), energy and environmental issues, partially as results of the emerging markets, problems deriving from the rising cost and declining quality of education, environmental regulation, demographic trends such as the ageing of the population in Western societies, rising tax burdens as well as massive consumer and government debt, which have been used to justify and impose austerity measures on whole countries such as Greece. Fuelled by consumption and government debt, consumption grew faster than real GDP over the last 40 years. However, Gordon (2012: 20) reckons that over a "substantial number of years in the future consumption must grow more slowly than production". Gordon predicts that "future growth in real GDP per capita will be slower than in any extended period since the late nineteenth century, and growth in real consumption per capita for the bottom 99% of the income distribution will be even slower than that" (Gordon 2012: 2). If, conversely, just 1% benefit from recent economic development, it is difficult to see which societal strata are supposed to carry the next economic upswing required for new growth.

Conclusion

This chapter has reviewed the socio-economic processes during which economic growth and the "growth paradigm" have become universal. While economic growth did not play any major role prior to the industrial revolution, the imperative for the economy to expand is inherent to more developed capitalism and anchored in its social relations and corresponding mindsets. In the most recent decades, economic growth was accompanied by an unbalancing between "real" and financial economy, unprecedented private and public debt, massively rising inequalities as well as an exacerbating ecological crisis (Chap. 4), which together have the potential of severely undermining the structural prospects for further growth. While the growth period that began in 1820 may be coming to a close, the growth period of the post-war decades took the form of a parallel advancement of profits and wages, and this was the structural precondition for the introduction and expansion of various welfare systems to which we turn next.

Notes

1. Moore (2015) refers to the structural preconditions necessary for long-term capital accumulation in terms of "four cheaps": labour power, food, energy and raw materials.
2. Crisis factors included the exhaustion of the productivity growth potentials of "scientific" management strategies, limits to product standardisation,

changes in product demand structures and in the international regulation of Fordism as well as an increased questioning of Fordism's male-breadwinner-based mode of societalisation and its fossil energy regime (Koch 2012; Bieling et al. 2016).

References

Agnew, J. 1987. *The United States in the World Economy*. London: Cambridge University Press.

Bieling, H.J., J. Jäger, and M. Ryner. 2016. Regulation Theory and the Political Economy of the European Union. *Journal of Common Market Studies* 54 (1): 53–69.

Boyer, R. 2000. Is a Finance-Led Growth Regime a Viable Alternative to Fordism? A Preliminary Analysis. *Economy and Society* 29 (1): 111–145.

Boyer, R., and Y. Saillard (eds.). 2002. *Régulation Theory. The State of the Art*. London: Taylor & Francis.

Burkett, P. 1999. *Marx and Nature. A Red and Green Perspective*. New York: St. Martin's Press.

Cleveland, C.J. 1999. Biophysical Economics: From Physiocracy to Ecological Economics and Industrial Ecology. In *Bioeconomics and Sustainability. Essays in Honor of Georgescu-Roegen*, ed. K. Mayumi and J.M. Gowdy, 125–154. Cheltenham: Edward Elgar.

Dale, G. 2012a. The Growth Paradigm: A Critique. *International Socialism* 134. http://isj.org.uk/the-growth-paradigm-a-critique/.

Dale, G. 2012b. Critiques of Growth in Classical Political Economy: Mill's Stationary State and a Marxian Response. *New Political Economy* 18 (3): 431–457.

Dale, G. 2015. Origins and Delusions of Green Growth. *International Socialist Review* 97. http://isreview.org/issue/97/origins-and-delusions-green-growth.

Dale, M., S. Krumdieck, and P. Bodger. 2011. Net Energy Yield from Production of Conventional Oil. *Energy Policy* 39 (11): 7095–7102.

Daly, H. 1972. In Defense of a Steady-State Economy. *American Journal of Agricultural Economy* 54 (5): 945–954.

Elchardus, M. 2011. Diversity and Standardization: Concepts, Issues and Approaches. In *Diversity, Standardization and Social Transformation: Gender, Ethnicity and Inequality in Europe*, ed. M. Koch, L. McMillan, and B. Peper, 9–26. London: Routledge.

Gordon, R.J. 2012. Is US Economic Growth Over? Faltering Innovation Confronts the Six Headwinds. Working Paper 18315, National Bureau of Economic Research, Cambridge.

Gramsci, A. 1971. *Selections from the Prison Notebooks*. Basingstoke: Macmillan.

Harvey, D. 1990. *The Condition of Postmodernity: An Enquiry into the Origins of Cultural Change*. Cambridge: Blackwell.

Jackson, T. 2009. Prosperity without Growth? The Transition to a Sustainable Economy. London: Sustainable Development Commission.

Jevons, W. S. 1865. *The Coal Question. An Inquiry Concerning the Progress of the Nation, and the Probable Exhaustion of Our Coal-Mines*. London: Macmillan.

Koch, M. 2006. *Roads to post-fordism. Labour markets and social structures in Europe*. 2nd ed. 2017. London: Routledge.

Koch, M. 2012. *Capitalism and Climate Change. Theoretical Discussion, Historical Development and Policy Responses*. Basingstoke: Palgrave Macmillan.

Koch, M. 2013. Employment Standards in Transition: From Fordism to Finance-Driven Capitalism. In *Non-standard Employment in Europe. Paradigms, Prevalence and Policy Responses*, ed. M. Koch and M. Fritz, 29–45. Basingstoke: Palgrave Macmillan.

Krugman, P. 2009. *The Return of Depression Economics and the Crisis of 2008*. New York: W.W. Norton & Company.

Maddison, A. 2007. *Contours of the World Economy, 1–2030 AD*. Oxford: Oxford University Press.

Marx, K. 1961. *Capital: A Critique of Political Economy*, vol. 1. Moscow: Foreign Languages Publishing House.

Marx, K. 2006. *Capital: A Critique of Political Economy*, vol. 3. London: Penguin Classics.

McNeill, J.R. 2000. *Something New Under the Sun: An Environmental History of the Twentieth-Century World*. New York: W.W. Norton & Company.

Mill, J.S. 1848. *Principles of Political Economy with Some of Their Applications to Social Philosophy*. London: John. W. Parker.

Moore, J.W. 2015. *Capitalism in the Web of Life: Ecology and the Accumulation of Capital*. London: Verso.

O'Neill, D. 2013. Gross Domestic Product. In *Degrowth: A Vocabulary for a New Era*, ed. G. D'Alisa, F. Demaria, and G. Kallis, 103–106. London: Routledge.

OECD. 2008. *Statistics, Knowledge and Policy: Measuring and Fostering the Progress of Societies*. Paris: OECD.

Polanyi, K. 1944. *The Great Transformation. The Political and Economic Origins of Our Time*. Boston: Beacon Press.

Smith, A. 1993. *An Inquiry into the Nature and Causes of the Wealth of Nations*. Oxford: Oxford University Press.

Soper, K., and M. Emmelin. 2016. Reconceptualising Prosperity: Some Reflections on the Impact of Globalisation on Health and Welfare. In *Sustainability and the Political Economy of Welfare*, ed. M. Koch and O. Mont, 44–58. London: Routledge.

Stiglitz, J. 2010. *Freefall. Free Markets and the Sinking of the Global Economy.* London: Penguin Books.

Stockhammer, E. 2008. Some Stylized Facts on the Finance-Dominated Accumulation Regime. *Competition and Change* 12 (2): 184–202.

UNCTAD. 2009. *World Investment Report.* New York: United Nations Publication.

Weber, M. 1958. *The Protestant Ethic and the Spirit of Capitalism.* London: Routledge.

CHAPTER 3

Growth and the Welfare State as We Know It

Abstract This chapter turns to the discussion of the evolution of welfare states in Western capitalism—one of the (bundles of) institutions that have become so relevant for supporting many aspects of people's well-being. It demonstrates the close interdependencies between growth and the welfare state. Economic growth, especially after World War II, freed up resources to finance welfare states and facilitate greater "social peace". Welfare state finances depend on a growth-generating economy to be sustained since taxes and social security contributions rise and fall with economic up- and downturns. The chapter also reviews the relationship between welfare regimes and environmental protection but finds that growth remains to be the main driver of environmental impacts.

Keywords Welfare state · Welfare regimes · Growth · Environmental policies

Chapter 2 demonstrated that in the post-war period, growth in productivity and GDP was associated with the achievement of economies of scale. This was a prerequisite for the simultaneous and proportionate development of production and consumption. Higher productivity decreased the percentage of wages of total employers' costs, while the real wages of workers increased at the same time because of falling prices. Employment was able to grow since the total volume of capital rose by a greater proportion than the increase in the number of workers made

M. Büchs and M. Koch, *Postgrowth and Wellbeing*,
DOI 10.1007/978-3-319-59903-8_3

redundant due to productivity gains in the work process. The cheapening of industrial products raised the purchasing power of wage labourers, so that both employers' profits and employees' real wages increased. The state benefited from this favourable situation and used its growing income from taxation for the expansion of welfare systems, which, amongst other things, guaranteed a minimum standard of living for those who did not participate in the labour market. This chapter first discusses the role of the state in producing economic growth in capitalist societies. It then focuses on the welfare state and welfare "regimes" and raises the issues of how these are linked to the pursuit of economic growth as well as how they typically perform in terms of inequality and the environment. Finally, we refer to current rescaling and recalibration trends within real-existing welfare states and ask, in anticipation of Chap. 7, what it would take to make welfare (systems) sustainable.

The State and the Pursuit of Growth in Capitalist Societies

In order to exchange goods as commodities, individuals must "recognize one another reciprocally as proprietors" (Marx 1973: 243). This includes a "juridicial moment" since exchange relations are only possible if the acting individuals are not prevented from entering them, for example, by feudal rule. Since the use of force is equally not a legal or legitimate course of action, the respect of the principle of equivalence in exchange relations depends on a formally independent institution that guarantees the legal and economic independence of the owners of commodities: their equality, legal security and protection. In the case of an advanced division of labour, this guarantee cannot be ensured in accordance with common law but must be institutionalised through an independent third party that, above all, monopolises the legitimate use of physical force (Weber 1991): the modern state. Hence, under the rule of law, an important role of the state in capitalist development is to guarantee private property, the principle of equivalence and the legal security of economic subjects.

Exchange relations, however, are not reduced to the swapping of use values. They also reproduce social relationships, which involve power asymmetries and social inequalities. The latter originate in different societal domains and take the form of class, race, religion, linguistic or gender characteristics. In a social structure based on a dynamic plurality of exploitative and exclusionary relationships (Koch 2006: 13–16), the state is the main location for the political regulation of conflicts and for the maintenance of social order (Offe 1984). Since, without state regulation, such a society would disintegrate, a further general role of the state is that of an arbiter to maintain a minimum of social cohesion and, at the same time, to legitimise remaining inequalities. In doing so, the state has an indispensable capability of temporarily harmonising conflicting group interests and creating consensus. A key issue here is the degree and the kind of commodification of socio-economic relations. It is far from being taken for granted and indeed a controversial matter what kind of private and social services and use values should take commodity form and be traded on markets which states often have to regulate. Child- and old-age care, education, the regulation of prostitution as well as of carbon emissions are prominent examples for this conflict. The state appears here as an autonomous political sphere, where social classes and groups represent their interests in indirect and mediated ways. As political parties and interest groups raise variable issues such as religion, age and the environment, these interests and issues are sometimes in the focus of government action, only to be superseded by others at later points in time. Hence, state policies cannot be reduced to the strategic interests of single actors, but rather develop as a result of the heterogeneity and changing dynamic of social forces that influence state institutions (Poulantzas 1978).

The historical development of markets and capital tends to dissolve previously isolated communities and to regroup their inhabitants according to new spatial-temporal structures. In most parts of Europe in the nineteenth and twentieth centuries, these largely followed the borders of the developing nation-states. Towards the end of the twentieth century, however, this particular state territoriality began to be faced with Europeanisation and internationalisation but also with localisation processes that undermined this spatial arrangement. As a consequence, scholars ceased to presuppose a static concurrence of nation and state and began to view state spatiality in more dynamic ways. The spatial dimension of state regulation is subject to rescaling processes in the course of which new, multi-scalar structures of state organisation, political authority and socio-economic regulation emerge (Koch 2008; Kazepov 2010). State institutions are central in what Brenner (2004: 453) describes as attempts of "spatial targeting": to "enhance territorially specific locational assets, to accelerate the circulation of capital, to reproduce the labour force, to address place-specific socio-economic problems and/or to maintain territorial cohesion". Similarly, the notion of "spatial-temporal fixes" (Harvey 2003) has been developed to reflect the fact that particular growth regimes correspond with particular scales of regulation or spatial boundaries (national, transnational, local) in which structural coherence is sought. Spatial-temporal fixes are associated with policy frameworks that target specific jurisdictions, places and scales as focal points for state regulation in particular periods of time.

Welfare States and Welfare Regimes

Chapter 2 demonstrated that the working class was actively integrated into the growth project of Fordism via the "management–labour balance" (Stockhammer 2008). The huge profit rates and real income increases, which fuelled the enormous economic growth rates in the period 1950–1975 in most Western countries, were the structural background for the development of modern welfare states and their various departments. Those who for different reasons did not work—for example the unemployed, sick, pensioners and students—received independent incomes, which the state raised via taxation from the primary incomes of employers and employees and subsequently redistributed to these groups. Welfare institutions provided not only safety nets against poverty but also protection against the loss of income in cases of unemployment, illness, disability or old age. Significant progress was achieved in population health and life expectancy as well as in material living standards.

If Marx succeeded in demonstrating that in capitalism labour products tend to take the form of commodities—and this includes the labour power which the majority of people need to sell on labour markets—Esping-Andersen's attempt to comparatively analyse real-existing welfare states focused on the kind and extent to which the (welfare) state "decommodifies" socio-economic relations by providing institutional protection of workers from total dependence for survival on employers (Esping-Andersen 1990). Welfare "regimes" take different forms and vary, above all, in terms of the particular division of labour of private and public provision to which different forms and extents of taxation correspond. In an elaboration of Korpi's (1983) power resources approach, Esping-Andersen argued that the extent and further particulars of decommodification processes depend on the structural strength of the Left vis-à-vis other political and economic forces. While a weak Left led to the development of liberal welfare states, especially in the Anglo-Saxon countries, strong conservative parties led to conservative welfare states in continental Europe. Finally, relatively strong Left parties supported the establishment of social-democratic welfare state, especially in Scandinavia.

These three welfare state types were associated with different patterns of stratification and the public–private mix of welfare. The "social-democratic" countries (for example Sweden and Norway) were characterised as having the highest degree of universalism in welfare programmes and offered the greatest redistribution and decommodification potential, including a weak role of the market in care for children and the elderly and, consequently, through the lowest degree of stratification; "liberal countries" such as the US, the UK or Ireland represent the lowest decommodification and the highest stratification potential and a corresponding emphasis on individual responsibility in combination with a greater welfare role of private agencies, charities, churches, employers and unions. Here, the primary goal of public welfare is not redistribution but the restoration of individuals' and families' self-sufficiency; in conservative countries with a Bismarckian welfare tradition (for example Germany or Switzerland) with medium decommodification and stratification state welfare policies, unemployment and pension policies, in particular, were not designed to redistribute market inequalities but tended to confirm and reinforce class differences in unemployment and pension benefits. The conservative welfare regime also tended to reproduce the traditional "male-breadwinner" model the most.

While different welfare regimes are associated with different results in terms of stratification and redistribution, there is no clear-cut answer as to whether certain welfare regimes are more dependent on the provision of economic growth than others. The further distinction between contribution-based and tax-based funding of welfare programmes appears to give ground to the argument that while contribution-based systems may react more directly than tax-based ones to market swings, the latter are more susceptible to state retrenchment than the former as legislators can reduce welfare expenditure when tax revenues become scarce. While an empirical test of such proposals is beyond the scope of the present volume, we suggest for the time being that state welfare expenditure in all welfare regimes remains dependent on the capability of taxing and redistributing the primary incomes of employers and employees—and hence on economic growth.[1] State taxing and spending was relatively easy in the post-war era where capital was normally spatially fixed in specific locations for long time periods and where the outsourcing of production tasks or entire production sites was much more difficult than today. As a proxy, (welfare) state revenues could grow with GDP. Conversely, welfare states played a significant part in stabilising and promoting growth, for example, through the provision of health insurance and health care, as well as training and education and most importantly perhaps through the introduction and expansion of unemployment insurance and minimum income schemes that contributed to keep up consumer demand during periods of unemployment.

The overall echo of the *Three Worlds of Welfare Capitalism* was very positive and long-lasting as a recent review of its reception history confirms (Emmenegger et al. 2015). The book has also stimulated "regime" and "typology debates" in various areas including care (Antonnen and Sipilä 1996; Simonazzi 2008), social assistance and minimum income (Gough et al. 1997), where similarities to and difficulties of welfare regime typologies have been highlighted. The overall positive reception is reflected in the fact that later welfare regime typologies that included greater numbers of countries confirmed rather than falsified Esping-Andersen's approach, insofar as they proposed four or five "worlds of welfare" rather than three, yet with significant overlap in the allocation of countries (Arts and Gelissen 2002; Ferragina and Seeleib-Kaiser 2011). In their influential overview of welfare regime typologies, Arts and Gelissen point out that the remarkable degree of theoretical consistency that characterises Esping-Andersen's approach would decrease through the adaptation of alternative theoretical arguments. However, beyond Esping-Andersen's classical types and in relation to Europe, two complementary welfare clusters are often distinguished to broaden the empirical reach of the comparative analysis: A "Mediterranean" cluster (e.g. Spain, Portugal and Greece) that, according to Ferrara (1996), also includes Italy, and a cluster of "Eastern European" countries, whereby the jury regarding the welfare affiliation of single Eastern European countries is still out (Fenger 2007). Similar debates are ongoing on other parts of the world including East Asia (Sung and Pascall 2014).

Welfare States and "Green" States

Most recently, Esping-Andersen's welfare regime approach has inspired debates on the so-called green or eco-social state (Gough et al. 2008; Koch and Fritz 2014) and, in response to the environmental crisis, between green growth and "no-growth" approaches (Khan and Clark 2016). According to researchers such as Dryzek et al. (2003) and Gough et al. (2008), social-democratic welfare states are better placed to manage the intersection of social and environmental policies than more liberal market economies and welfare regimes. One reason Dryzek mentions is the discourse on "ecological modernisation", which he regards as especially widespread in the Nordic countries: the idea that environmental policies can be good for business and that "green growth" presupposes the governance capacities of coordinated markets. Rather than trusting the invisible hand of the market, social-democratic welfare regimes would generally make a "conscious and coordinated effort and regard economic and ecological values as mutually reinforcing" (Gough et al. 2008: 334–335). The "contemporary result" would be the "mainstreaming of both environmental and equality concerns" (Gough et al. 2008: 330). Similarly, Meadowcroft (2005) and Gough et al. (2008) argue that there are a range of linkages between social and environmental policies which together have the potential of bringing about sustainable development.

While there seem to be good theoretical reasons to assume that social-democratic welfare regimes provide a better institutional basis for the introduction and development of the "green" dimension of the state than conservative and, especially, liberal welfare regimes, all mentioned authors are in agreement that this institutional basis is no guarantee that green states de facto develop in synergy with the welfare state. In fact, all mention the possibility of competition, clashes and conflicts between the two. This possibility is even more emphasised by ecological economists such as Victor (2008); Jackson (2009) and Daly and Farley (2011), who question both the synergy hypothesis of the welfare and green dimension of the state and the "green growth" policy option that follows from it (Koch 2013). Instead, these authors regard both welfare and environmental performance of a country primarily as a reflection of its development in economic terms, that is, of GDP growth. Hence, while "green growth" and "ecological modernisation" discourses claim that the pursuit of economic growth can be made compatible with sustainable development targets by building on existing (welfare) institutions, post-growth theories and the mentioned ecological economists would regard economic growth itself as the problem (see also Chap. 4). Accordingly, GDP growth would need to be deprioritised in policy making across the advanced capitalist world—that is, irrespective of welfare affiliation—in order to allow for efficient environmental policy making and achieve ecological sustainability.

The claim that social-democratic welfare regimes, which are least unequal in socio-economic terms, would also perform best in ecological and climate terms (Gough et al. 2008) and gradually turn into "eco-social states" could not be verified in comparative empirical research. Instead, representatives of all welfare regimes—social-democratic, conservative and liberal—are to be found amongst relatively well, medium and poor performing "green" states (Koch and Fritz 2014). Contradicting the "synergy hypothesis", the paradox of the Western welfare state seems to lie in the fact that the same mechanisms that defuse socio-economic inequalities ensure the inclusion of an increasing amount of people in environmentally problematic production and consumption practices. While existing welfare states contribute to the generation of wellbeing, partly due to its decommodification effect on market mechanisms, this is coupled to the growth paradigm, thereby complicating the promotion of alternative wellbeing conceptions and practices. Indeed, rather than welfare regimes, the level of economic development measured in GDP per capita turned out to be most responsible for countries' ecological (under-) performance. Such recent comparative empirical results largely confirm previous studies that fundamentally question "green growth" policy options—the idea that economic growth can be organised in both socially equitable and ecologically sustainable ways (Victor 2008; Jackson 2009). In fact, if welfare goals are to be combined with ecological sustainability, let alone generalised to all inhabitants of the globe, it is difficult to see how the top priority of economic growth in policy making can continue.

Recalibration and Rescaling: Towards Sustainable Welfare States?

Recent welfare literature has highlighted recalibration and rescaling trends of national welfare states as they developed in the postgrowth circumstances. Both trends should be seen against the background of a much changed global political economy where capital is more geographically mobile than in Fordism and where it has, mainly due to developments in information and communication technologies, become much easier to outsource parts of the production process so that planning, production and sales of commodities often take place in different parts of the world. State tax and spend strategies have become accordingly more difficult, since world locations increasingly compete with each other to attract Foreign Direct Investment, especially financial capital. Welfare state *recalibration* trends suggest a transfer or responsibilities from the state to individuals, markets and the non-profit sector (as well as reorganising welfare spending priorities within the state). Many Western states have pushed through welfare reforms that place stronger emphasis on citizens' duties and responsibilities. This includes changes in income protection systems in terms of access, decreasing benefit levels and increasing reliance on means-tested benefits (Angelin et al. 2014). Individuals are meant to be financially self-sufficient, and this includes that they rely on other income sources than the welfare state, including loans from families and financial markets. The welfare state has largely withdrawn from areas such as subsidised rental housing and comprehensive education, promoting home ownership and loan-financed education in the semi-private or private sector instead. In addition, governments have ceased to fund longer periods of unemployment and encourage individuals to set up small businesses or become self-employed. These trends have gone furthest in Anglo-Saxon countries but are also evident in other welfare systems including the social-democratic ones (for the Swedish experience, see Koch 2016). The result is a revised social contract between the individual citizen and the state, where the post-war balance of entitlements and duties in the provision of welfare has been transferred so that it is now individual citizens who are expected to take responsibilities for these welfare concerns—and to pay for them (Crouch 2009).

Welfare state recalibration trends are accompanied by *rescaling* processes. Core responsibilities of the originally *national* welfare state have been delegated upwards to the European level and downwards to local levels. Many welfare targets such as unemployment or poverty rates for different social groups are now set at the European level. To meet them, the European Union (EU) applies the Open Method of Coordination, an iterative procedure, involving a rolling programme of annual planning, monitoring, examining and readjusting whereby national welfare policies are put to the test of cross-country comparison, including peer-review (Büchs 2007). A range of quantitative benchmarks are being specified that EU member states are supposed to achieve in National Reform Programmes. At the same time, many EU member states have shifted welfare responsibilities downwards to local levels. This is especially evident in the area of minimum income protection and social assistance (Johansson and Panican 2016).

In a situation where the welfare state "as we know it" is undergoing recalibration and rescaling processes anyway, some researchers have started to also address environmental issues in this context and ask how the meaning of welfare changes if we take environmental sustainability seriously and how this could be provided at multiple scales (Koch and Mont 2016). What Langhelle (1999) calls the "sustainability proviso" means a widened scope and changed pattern of welfare provision and, particularly, of any distributive principle applied. First, one would need to take into account not only a given (welfare) state constituency but also non-citizens, even if the governance of a given nation-state does not formally include temporally and geographically distant people (Brandstedt and Emmelin 2016). Second, any sustainable welfare provision would also need to consider that the satisfaction of present needs and wants must not compromise or undermine the ability of future generations' needs satisfaction. This presupposes the recognition of limitations. Third, central to a concept of sustainable welfare is the emphasis on human needs and universalisability. The main welfare concern is not the unlimited provision with material riches of the happy few in Western societies but the satisfaction of basic needs for all humans now and in the future. We will return to the concept of human need and sustainable welfare in Chaps. 5 and 7.

Conclusion

The state plays indispensable roles in capitalist development as well as in the provision of economic growth and social cohesion. Welfare states counteract the capitalist dynamic in "decommodifying" social relationships, especially the obligation to sell one's labour power for a living. Extent and ways in which this process proceeds, vary from country to country and according to welfare regimes. However, hopes that social-democratic welfare regimes would not only be most redistributive but also best-performing in terms of ecological sustainability could not be verified in comparative research. Environmental damages such as carbon emissions and too large ecological footprints are instead largely associated with the level of economic development measured as GDP per capita. Since an embedding of economy and society in environmental limits is unlikely to happen if the top priority of economic growth in policy making is upheld, the next chapters will discuss the alternative of providing welfare and wellbeing in non-growing economies.

Note

1. Bailey (2015) discusses the considerable fiscal difficulties that existng welfare state systems would face in the absence of growth.

References

Angelin, A., H. Johansson, and M. Koch. 2014. Patterns of Institutional Change in Minimum Income Protection in Sweden and Germany. *Journal of International and Comparative Social Policy* 30 (2): 165–179.

Antonnen, A., and J. Sipilä. 1996. European Social Care Services: Is it Possible to Identify Models? *Journal of European Social Policy* 6 (2): 87–100.

Arts, W., and J. Gelissen. 2002. Three Worlds of Welfare Capitalism or More? A State-of-the-Art Report. *Journal of European Social Policy* 12 (2): 137–158.

Bailey, D. 2015. The Environmental Paradox of the Welfare State: The Dynamics of Sustainability. *New Political Economy* 20 (6): 793–811.

Brandstedt, E., and M. Emmelin. 2016. The Concept of Sustainable Welfare. In *Sustainability and the Political Economy of Welfare*, ed. M. Koch, and O. Mont, 15–28. London: Routledge.

Brenner, N. 2004. Urban Governance and the Production of New State Spaces in Western Europe, 1960–2000. *Review of International Political Economy* 11 (3): 447–488.

Büchs, M. 2007. *New Governance in European Social Policy. The Open Method of Coordination*. Basingstoke: Palgrave Macmillan.

Crouch, C. 2009. Privatised Keynesianism: An Unacknowledged Policy Regime. *The British Journal of Politics and International Relations* 11 (3): 382–399.

Daly, H., and J. Farley. 2011. *Ecological Economics. Principles and Applications*, 3rd ed. Washington: Island Press.

Dryzek, J., D. Downes, C. Hunhold, D. Schlosberg, and H. Hernes. 2003. *Green States and Social Movements: Environmentalism in the United States, United Kingdom, Germany and Norway*. Oxford: Oxford University Press.

Emmenegger, P., Kvist, J., Marx, P. and Petersen, K. 2015. 'Three Worlds of Welfare Capitalism: The Making of a Classic'. *Journal of European Social Policy* 25 (1): 3-10.

Esping-Andersen, G. 1990. *The Three Worlds of Welfare Capitalism*. Cambridge: Polity Press.

Fenger, H. 2007. Welfare Regimes in Central and Eastern Europe: Incorporating Post-Communist Countries in a Welfare Regime Typology. *Contemporary Issues and Ideas in Social Sciences* 3 (2): 1–30.

Ferragina, W., and M. Seeleib-Kaiser. 2011. Thematic Review: Welfare Regime Debate—Past, Present, Futures? *Policy and Politics* 39 (4): 583–611.

Ferrara, M. 1996. The 'Southern' Model of Welfare in Social Europe. *Journal of European Social Policy* 6 (1): 17–37.

Gough, I., J. Bradshaw, J. Ditch, T. Eardley, and P. Whiteford. 1997. Social Assistance in OECD Countries. *Journal of European Social Policy* 7 (1): 17–43.

Gough, I., J. Meadowcroft, J. Dryzek, J. Gerhards, H. Lengfield, A. Markandya, and R. Ortiz. 2008. JESP Symposium: Climate Change and Social Policy. *Journal of European Social Policy* 18 (4): 25–44.

Harvey, D. 2003. *The New Imperialism*. Oxford: Oxford University Press.

Johansson, H., and A. Panican (eds.). 2016. *Combating Poverty in Local Welfare Systems*. Basingstoke: Palgrave Macmillan.

Jackson, T. 2009. Prosperity without Growth? The Transition to a Sustainable Economy. London: Sustainable Development Commission.

Kazepov, Y. (ed.). 2010. *Rescaling Social Policies towards Multilevel Governance in Europe*. Aldershot: Ashgate.

Khan, J., and E. Clark. 2016. Green Political Economy. Policies for and Obstacles to Sustainable Welfare. In *Sustainability and the Political Economy of Welfare*, ed. M. Koch, and O. Mont, 77–93. London: Routledge.

Koch, M. 2006. *Roads to post-fordism. Labour markets and social structures in Europe*. 2nd ed. 2017. London: Routledge.

Koch, M. 2008. The State in European Employment Regulation. *Journal of European Integration* 30 (2): 255–272.

Koch, M. 2013. Welfare after Growth: Theoretical Discussion and Policy Implications. *International Journal of Social Quality* 3 (1): 4–20.

Koch, M. 2016. The Role of the State in Employment and Welfare Regulation: Sweden in the European Context. *International Review of Social History* 61 (S24): 243–262.

Koch, M., and M. Fritz. 2014. Building the Eco-Social State: Do Welfare Regimes Matter? *Journal of Social Policy* 43 (4): 679–703.

Koch, M., and O. Mont (eds.). 2016. *Sustainability and the Political Economy of Welfare*. London: Routledge.

Korpi, W. 1983. *The Democratic Class Struggle*. London: Routledge & Kegan Paul.

Langhelle, O. 1999. Sustainable Development: Exploring the Ethics of Our Common Future. *International Political Science Review* 20 (2): 129–149.

Marx, K. 1973. *Grundrisse: Foundations of the Critique of Political Economy*. Harmondsworth: Penguin.

Meadowcroft, J. 2005. From Welfare State to Ecostate. In *The State and the Global Ecological Crisis*, ed. J. Barry, and R. Eckersley, 3–23. Cambridge: MIT Press.

Offe, C. 1984. *Contradictions of the Welfare State*. ed. John Keane. Cambridge, MA: MIT Press.

Poulantzas, N. 1978. *State, Power, Socialism*. London: NLB.

Simonazzi, A. 2008. Care Regimes and National Employment Models. *Cambridge Journal of Economics* 33 (2): 211–232.

Stockhammer, E. 2008. Some Stylized Facts on the Finance-Dominated Accumulation Regime. *Competition and Change* 12 (2): 184–202.

Sung, S., and G. Pascall (eds.). 2014. *Gender and the Welfare States in East Asia*. Basingstoke: Palgrave Macmillan.

Victor, P.A. 2008. *Managing Without Growth: Slower by Design, Not Disaster*. Cheltenham: Edward Elgar.

Weber, M. 1991. Politics as a Vocation. In *From Max Weber: Essays in Sociology*, ed. H. Gerth, and C. Wright Mills, 77–128. London: Routledge.

CHAPTER 4

Critiques of Growth

Abstract This chapter reviews the debates that highlight the problematic ecological and social consequences of growth. It argues that more recent limits to growth positions have moved away from a focus on resource limitations. Instead, concerns about the surpassing of a range of "planetary boundaries", especially potentially catastrophic and irreversible effects of climate change, have become more important for this debate, as it is already affecting people's wellbeing. The final part of this chapter introduces ideas of postgrowth. It provides an overview of different positions within this field, distinguishing system-reform, anti-capitalist and alternative-open approaches which differ in the ways in which they criticise growth and consider its relationship to growth.

Keywords Limits to growth · Climate change · Planetary boundaries Resources · Postgrowth

As the previous chapters have shown, economic growth is regarded as a prime policy aim by policy makers and economists because it is thought to be essential for reducing poverty and generating rising living standards and stable levels of employment (Ben-Ami 2010: 19–20). More generally, support for economic growth is usually intertwined with advocating social progress based on scientific rationality and reason and hence with an optimistic view of humans' ingenuity to solve problems (ibid.: 17, 20, Chap. 5). Growth criticism thus tends to be portrayed as anti-progress

M. Büchs and M. Koch, *Postgrowth and Wellbeing*,
DOI 10.1007/978-3-319-59903-8_4

and inherently conservative (ibid.: Chap. 8). While it is important to acknowledge and discuss this view, it needs to be emphasised that growth criticism is formulated with long-term human welfare in mind which advocates alternative types of social progress (Barry 1998). This chapter first outlines ecological and social strands of growth critiques and then introduces relevant concepts of and positions within the postgrowth debate.

Ecological Critiques of Growth

Generally speaking, two types of growth criticism can be distinguished: the first focuses on limitations of GDP as a measure of economic performance; the second goes beyond this by highlighting the inappropriateness of growth as the ultimate goal of economic activity and its negative implications for environment and society.

Since GDP measures the monetary value of all final goods and services in an economy, it excludes the environmental costs generated by production. For instance, as long as there is no cost associated with emitting greenhouse gases , the cost for the environmental and social damage following from this is not reflected in GDP figures. Worse even, GDP increases as a consequence of some types of environmental damage: if deforestation and timber trade increase or if natural disasters or industrial accidents require expenditures for clean-up and reconstruction, GDP figures will rise (Douthwaite 1999: 18; Leipert 1986). Several critics of GDP as a measure of progress have proposed alternative indicators of welfare such as the Genuine Progress Indicator, Green GDPs or other approaches which factor in environmental costs (see Chap. 5 for more details), but they do not necessarily object to economic growth being the primary goal of economic activity (van den Bergh 2011).

In contrast, the idea of ecological limits to growth goes beyond the critique of GDP as a measure of economic performance. Instead, it maintains that economic growth should not, and probably cannot, be the main goal of economic activity because it requires increasing resource inputs, some of which are non-renewable, and generates wastes, including greenhouse gases, that disturb various ecosystems, severely threatening human and planetary functioning in the short and long term.

Resources are regarded as non-renewable if they cannot be naturally replaced at the rate of consumption (Daly and Farley 2011: 75–76). Examples include fossil fuels, earth minerals and metals, and some nuclear materials like uranium (Daly and Farley 2011: 77; Meadows et al. 2004: 87–107). Based on work by Georgescu-Roegen (1971), many ecological economists also assume that non-renewable resources cannot be fully recycled because they become degraded in the process of economic activity. Historically speaking, economic growth is a fairly recent phenomenon (Fig. 2.1). Since its onset in the late seventeenth century in Europe and mid-eighteenth century in the US (Gordon 2012), it has gone hand in hand with an exponentially increasing use of non-renewable resources such as fossil fuels (Fig. 4.1).

While we are not yet close to running out of non-renewable resources, over time they will become more difficult and hence more

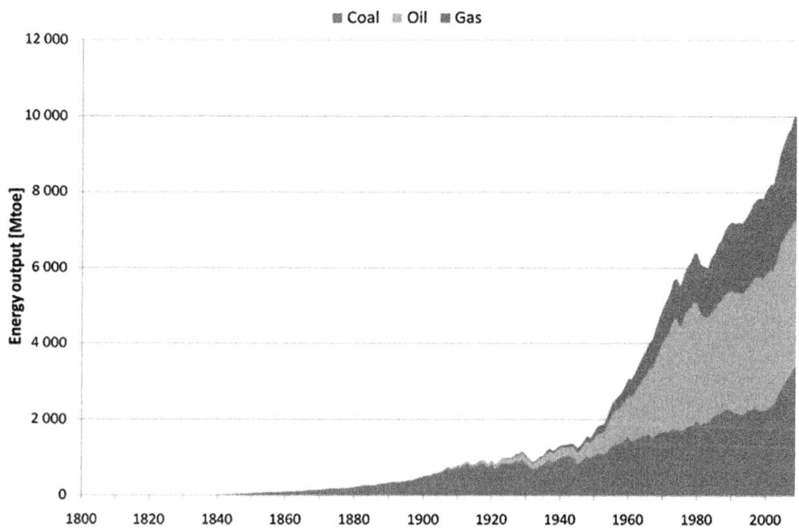

Fig. 4.1 Production of fossil energy in the world from 1800 to 2009. *Source* © 2011 Höök M. Fuelling future emissions—Examining fossil fuel production outlooks used in climate models. In: Blanco, J. and Kheradmand, H. eds. Climate change—Research and technology for adaptation and mitigation. In TechOpen, under CC BY-NC-SA 3.0 license. Available from: http://dx.doi.org/10.5772/24848

expensive to recover. This idea is captured by the concept of "energy returned on energy invested" (EROEI). In relation to oil for instance, it has been shown that the easily recoverable fields have been targeted first and that therefore greater energy (and hence financial) inputs will be required to produce more oil. Over time, the ratio of energy returned on energy invested will decrease, reducing the financial incentive to invest further in the recovery of these non-renewable resources (Dale et al. 2011; Brandt et al. 2015: 2). Relevant to this is also the debate about peak oil—a concept coined by Shell Oil geologist Marion King Hubbert in the 1950s—the point at which the rate of global conventional oil production reaches its maximum which is expected to take place roughly once half of global oil reserves have been produced. There is still controversy about whether global peak oil will occur, and if so when, as it is difficult to predict, or get reliable data on, the rate at which alternative types of energy will replace oil (if this was to happen fast enough, peak oil might not be reached, if it has not yet occurred), the size of remaining oil reserves and the future efficiency of oil extraction technologies (Chapman 2014). However, it is plausible to assume that oil prices will rise in the long term if conventional oil availability diminishes, while global demand for oil increases with continuing economic and population growth. Since economic growth in the second half of the twentieth century required increasing inputs of conventional oil, higher oil prices would have a negative impact on growth unless alternative technologies are developed that can generate equivalent liquid fuels at lower prices (Murphy and Hall 2011).

Some scholars have criticised the focus on physical/energy resource limitations as initially highlighted in the "limits to growth" debate (Meadows et al. 1972) and state that instead catastrophic climate change is likely to be a more serious and immanent threat to humanity (Schwartzman 2012). The main arguments here are first that much uncertainty remains about the potential and timing of peak oil, future availability of other fossil fuels and development of alternative low energy resources, while the impacts of climate change are already immanent and may accelerate within the very near future. Second, even if peaks in fossil fuel production occurred in the near future, remaining resources could still be exploited to their maximum. However, this would be devastating from a climate change perspective as, according to the latest IPCC scenarios, greenhouse gas emissions need to turn net-zero by the second half of this century for there to be a good chance to limit global warming to 2° Celsius (and ideally, below that) (Anderson and Peters 2016). It is telling that some of the more recent debates about ecological limits to growth put much more emphasis on environmental impacts of growth, rather than on peak oil or other resource limitations (Dietz and O'Neill 2013). Differently put, limits of sinks, especially to absorb greenhouse gases, and to the regeneration of vital ecosystems are now attracting greater concern, compared to limits of resources.

Growing economic production generates increasing pressures on the environment due to pollution of air, water and soil, the destruction of natural habitats and landscapes, for instance, through deforestation and the extraction of natural resources. Therefore, growth often also threatens the regeneration of renewable resources such as healthy soil, freshwater and forests, as well as the functioning of vital ecosystems and ecosystems services such as the purification of air and water, water absorption and storage and the related mitigation of droughts and floods, decomposition and detoxification and absorption of wastes, pollination and pest control (Meadows et al. 2004: 83–84). Recent research on planetary boundaries has started to identify thresholds of environmental pollution or disturbance of a range of ecosystems services beyond which the functioning of human life on earth will be put at risk. Rockström and colleagues have identified nine such "planetary boundaries"—"climate change; rate of biodiversity loss (terrestrial and marine); interference with the nitrogen and phosphorus cycles; stratospheric ozone depletion; ocean acidification; global freshwater use; change in land use; chemical pollution; and atmospheric aerosol loading" (Rockström et al. 2009: 472). They also present evidence according to which three of these boundaries—climate change, rate of biodiversity loss and the nitrogen cycle—have already reached their limits (Rockström et al. 2009).

Of those three thresholds, climate change has received most attention. The 5th Assessment Report of the Intergovernmental Panel on Climate Change (IPCC 2014) concluded that global temperatures have risen by an average of 0.85° since the 1880s (while local temperature increases can be much higher than that) and that the concentration of greenhouse gases in the atmosphere has reached unprecedented levels over the last 800,000 years—that of CO_2 has now reached 405.6 parts per million (NASA, January 2017, Fig. 4.2), far surpassing the level of 350 ppm which is considered safe by many scientists (Rockström et al. 2009). The IPCC report also maintained that humans very likely

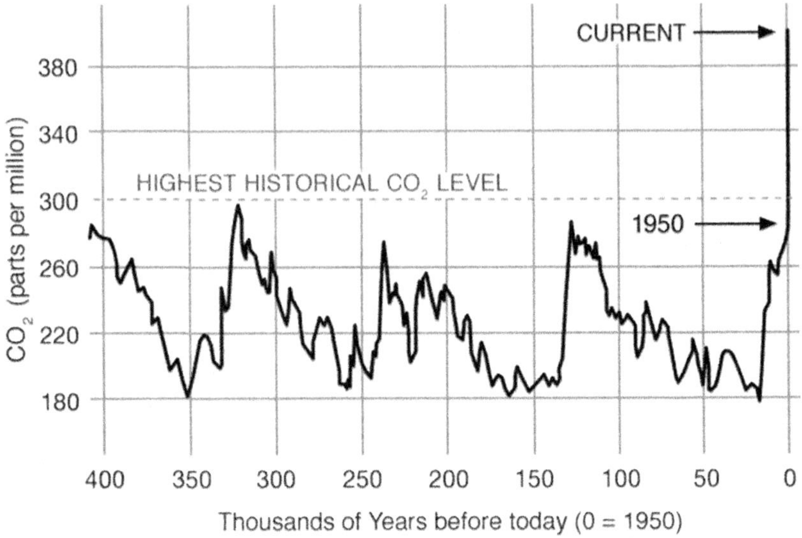

Fig. 4.2 Concentration of CO_2 in the atmosphere. *Source* NASA, available from https://climate.nasa.gov/vital-signs/carbon-dioxide/. The CO_2 levels have been reconstructed from measures of trapped air in polar cap ice cores

contributed to at least 50% of global warming that occurred since the 1950s (IPCC 2014: 5). A range of climate change impacts can already be observed, including a 26% increase of ocean acidification since industrialisation; shrinking of glaciers, Greenland and Antarctic ice sheets, as well as arctic sea ice; and the rise of sea levels of 19 cm since 1901. This is projected to increase by an additional 82 cm by the end of this century at current levels of greenhouse gas emissions (ibid.: 13). Climate change impacts are already felt with increased occurrences of heat waves, heavy rain fall, increased risk of flooding and impacts on food and water security in a number of regions around the world. It is projected that with a rise of 2° of global temperatures, 280 million people worldwide (with greatest numbers in China, India and Bangladesh) would be affected by sea level rise, escalating to a projected 627 million people under a 4° scenario (Strauss et al. 2015: 10).

At the 21st Conference of Parties of the United Nations Framework Convention on Climate Change in Paris in 2015, representatives agreed that action should be taken to limit rise of global temperatures to 2° and to "pursue efforts" to limit it to 1.5°. This has been adopted by 196 countries, but immense efforts and very radical reductions of greenhouse gas emissions will be required to comply with the agreement. Even if net greenhouse gas emissions were reduced to zero, surface temperatures would remain constant at their increased levels for hundreds of years to come and climate change impacts such as ocean acidification and rising sea levels would continue for hundreds or even thousands of years once global temperatures are stabilised; moreover, a range of climate change impacts are deemed irreversible (IPCC 2014: 16).

One controversial question in the debate about economic growth and environmental impacts has been whether growth can be decoupled from the damage it causes. Important to this debate is the theory of the Environmental Kuznets Curve which applies Simon Kuznets' hypothesised inverted u-shaped relationship between economic development and income inequality to the relationship between economic development and environmental degradation. According to this theory, environmental degradation is low in the early phases of economic development, then rises with increasing development up to a certain point, beyond which it falls again with advancing development because more resources can be invested to render production and consumption more efficient and less polluting. Therefore, this theory suggests that it is possible to decouple economic growth (measured in GDP) from its environmental implications.

The counter-argument to this theory is that it does not take into account the difference between relative and absolute decoupling. Relative decoupling refers to the environmental impacts generated over time per unit of economic output, for instance CO_2 emissions per million of US$. In contrast, absolute decoupling would examine aggregate environmental impact, compared to total economic output over time. Here it has been argued that while relative decoupling may be possible as the environmental impact *per unit* of economic output decreases over time due to efficiency gains, absolute decoupling is much harder to achieve while growth continues. Indeed, there is no evidence for absolute decoupling as total environmental impacts, for instance total global CO_2 emissions, are still rising with rising global GDP (Jackson 2011: 67–86). This is partly due to rebound effects which we discussed in Chap. 2: rising consumption because the increase in efficiency has made it cheaper to produce/consume (Jackson 2011: 67–86; see also Czech 2013: Chap. 8 criticising "green growth"). Furthermore, if decoupling is examined at the country level, one would need to take consumption-based resource use/emissions into account rather than production-based impacts. Substantial environmental impacts related to everything that is consumed in rich countries occur in developing countries from which goods are imported. A focus on production-based environmental impacts would hence be misleading as it ignores the environmental impacts that relate to a country's living standards and that occur outside of that country.

Social Critiques of Growth

Economic growth has not only been criticised from an ecological perspective, but also from an individual and social wellbeing point of view. Here, we can again distinguish a critique of GDP as a measure of wellbeing and a wider critique which highlights potential negative consequences of economic growth for human wellbeing.

Several scholars have argued that GDP is an inadequate measure of prosperity or wellbeing because it only includes market transactions and ignores activities of the informal economy in households and the volunteering sector which make an important contribution to individual and social wellbeing (Stiglitz et al. 2011; van den Bergh 2009; Jackson 2011). It also excludes the contribution of certain government services that are provided for free (Douthwaite 1999: 14; Stiglitz et al. 2011: 23), and the roles of capital stocks and of leisure in generating welfare (Costanza et al. 2015: 137). Furthermore, all market transactions make a positive contribution to GDP, regardless of whether expenditures increase or decrease welfare. Similar to the way in which environmental costs of growth are either excluded from GDP or even increase it, expenditures that arise from road accidents, divorces, crime, etc., contribute positively to GDP (ibid.: 133). The focus on market transactions also means that an increasing marketisation (or "commodification") of an economy will be reflected in a rise of GDP, which may or may not be related to actual "welfare" outcomes (Stiglitz et al. 2011: 49). It also implies that GDP is an insufficient cross-national comparator for the quality of life, as it does not take into account the different sizes of the informal economy across countries (ibid.: 15).

Furthermore, GDP does not indicate how income and consumption are distributed in society (Stiglitz et al. 2011: 44). This implies that a rise of GDP can be consistent with a rise of inequality of income and wealth.

However, if greater inequality has negative impacts on social wellbeing (Wilkinson and Pickett 2009), this would be masked by rising GDP figures (Douthwaite 1999: 17).

An even more fundamental criticism of GDP as a measure of wellbeing is that it focuses on the accumulation of money or wealth and thus on the material aspects of wellbeing. Such a narrow conception of the goals of economic activity and wellbeing has been criticised early on in the history of economic thought, e.g. by Aristotle's distinction between *oikonomia* and *chrematistics*. The latter refers to the accumulation of wealth and was regarded by him as an "unnatural" activity which did not contribute to the generation of use value and wellbeing (Cruz et al. 2009: 2021). The argument that wider conceptions of wellbeing and prosperity are required has also become relevant for contemporary critiques of economic growth (Jackson 2011; Paech 2013; Schneider et al. 2010) as we will discuss this in more detail in Chap. 5.

Arguments About the Psychological and Social Costs of Growth

The broader social critique of economic growth highlights potential "social limits" to or even negative consequences of economic growth for individual and collective wellbeing. The term "social limits to growth" was coined by Fred Hirsch (1976). He argued that the benefits of growth are initially exclusive to small elites and that these benefits disappear as soon as they spread more widely through mass consumption. For instance, only few people can own a Rembrandt painting; holiday destinations are more enjoyable when they are not overrun by hordes of other tourists; there are only few leadership positions, etc. From this perspective, there are "social limits" to the extent to which the benefits of growth can be socially expanded and equally shared.

Other scholars have expressed concern about individual and collective social costs of economic growth. First, there is the argument that the need to keep up with ever-rising living standards and new consumer habits, "keeping up with the Joneses"—a lot of which is seen to be driven by advertisement and social pressure rather than real needs, for instance fashionable clothing or gadgets—can generate stress and increase the occurrence of mental disorders (James 2007; Offer 2006; Kasser 2002).

Second, it has been argued that economic growth can imply wider social costs. For instance, with its emphasis on individual gain, market relations and competition, and the need that it generates for spatial mobility (e.g. for successful participation in education and labour markets), it is feared to undermine moral and social capital and put a strain on family and community relations, potentially even leading to increasing divorce and crime rates (Douthwaite 1999; Daly and Cobb 1989: 50–51; Hirsch 1976). Social costs of technological development and industrialisation also include industrial workplace and traffic accidents and time lost in traffic jams and for commuting (Czech 2013: Chap. 2; Stiglitz et al. 2011: 24). Technological innovation which arises from growth can also act as a factor for job losses and increasing job insecurity (Douthwaite 1999), especially if growth rates are not sufficiently high to compensate gains in productivity.

It is often assumed that growth will benefit the many because of assumed "trickle-down" effects which promise to improve the lot of the poor simply because the "cake" of available wealth is growing. While progress has been made in reducing extreme global poverty and inequality (Sala-i-Martin 2006; Rougoor and van Marrewijk 2015), the number of people living in poverty across the globe remains high.[1] At the same time, income inequality in a range of countries has been rising and the situation of many of the people living in extreme poverty is not improving which means the fruits of economic growth remain to be unequally distributed (Collier 2007; Piketty and Saez 2014).

The post-development debate goes even further than that in arguing that not only may growth not have reached the global poor to the extent that had been predicted by neoclassical economists, but that it can also have negative impacts on indigenous communities in developing countries, especially those who rely on local natural resources for their livelihoods which often suffer exploitation, pollution or even destruction through the inclusion of local economies into global value chains (Rahnema and Bawtree 1997).

While the distinction between critiques of growth that focus on its problematic ecological and social consequences is useful for analytic purposes, the two dimensions are of course closely linked. Ecological consequences of growth have the potential to severely impact or even undermine human wellbeing. Local livelihoods are already affected by current climate change impacts such as ocean acidification and its impact on marine organisms, draughts, floods and severe weather events, the frequency of which has been rising. Accordingly, it is estimated that crop and fish yields are already diminishing in several regions (Stern 2015; IPCC 2014) and that millions of people are already being displaced and forced to migrate due to climate change and other environmental impacts (Black et al. 2011). While the overall long-term impacts of climate change and the surpassing of other planetary boundaries are difficult to predict, they clearly have the potential to substantially undermine human wellbeing. Since greenhouse gas emissions are driven by economic growth, the development of alternative economic models that do not depend on growth is urgent since continued growth "threatens to alter the ability of the Earth to support life" (Daly and Farley 2011: 12).

Postgrowth: Concepts and Positions

Based on growth-critical discourses, numerous scholars have developed alternative concepts of the economy which aim for long-term environmental and social sustainability. Two main proposals can be identified here: the first is a steady-state economy (SEE) of constant (but internally dynamic) stocks of physical capital (artefacts) and population with a constant rate of throughput—the flow of raw materials from the environment through the economy and back to the environment as waste—that is sustainable in the long term (Daly and Farley 2011: 55–56). The second is degrowth which can be regarded as transitory phase of economic contraction in wealthy countries whose ecological footprints currently far exceed sustainable scales. In this book, we use the term "postgrowth" as an overarching term to capture these two dimensions which are compatible if degrowth is understood as a temporary process of reaching a sustainable steady state (Kerschner 2010).

As mentioned in Chap. 2, the formulation of the idea of an SEE is often (uncritically) attributed to John Stuart Mill, who referred to it as "stationary state" (Mill and Laughlin 1884). It is important to stress here that ecological economists do not regard a steady state as static. Instead, an SEE is open to development which is defined as "qualitative change (…) [and] evolution toward an improved but not larger structure or system" (Daly and Farley 2011: 6).

Underlying the vision of an SEE is the idea that the economy is a subsystem of the ecosystem, rather than the other way round or of them being separate systems. This means that the economic system is seen as dependent on the ecosystem and that economic activity needs to operate within certain limits that allow the ecosystem to function and to regenerate itself. This is where the definition of the "optimal scale" of the economy becomes relevant. Daly and Farley (2011: 20–21) provide the following abstract definition of optimal scale by applying the idea of marginal cost and utility from neoclassical micro-economics to macroeconomics. Accordingly, the optimal scale for the economy as a whole is the point where marginal utility from economic activity equals marginal disutility (e.g. pollution)—the economy should not grow beyond this point and this is the optimal scale at which the SEE should operate.

The invention of the term degrowth (in French—*décroissance*) in the early 1970s is often accredited to André Gorz (Kallis et al. 2015: 1). Degrowth supporters argue that a phase of economic contraction is required in wealthy countries to reach a sustainable scale of production and consumption. They also emphasise that degrowth is not equivalent to just shrinking GDP or to economic crisis. Instead, it is meant to be "planned" and democratically agreed upon processes which create a fundamentally different economic systems, centred around sharing, simplicity, care, conviviality and the commons (ibid.: 3).

A range of positions can be identified in the postgrowth debate. They differ in relation to the ways in which they criticise growth and consider its relationship to capitalism, and the visions they put forward for postgrowth societies. Broadly, we can distinguish the following positions: system-reform, anti-capitalist and alternative-open (e.g. Schmelzer 2015; Adler and Schachtschneider 2010).

System-reform approaches criticise GDP as a measure of economic welfare and social progress and argue it should be replaced with alternative goals that include social and environmental measures. According to this position, one can be agnostic about GDP growth once alternative policy goals have been formulated as growth may or may not contribute to their achievement. In other words, representatives of this position argue the focus does not need to be the reduction of GDP growth as long as a more holistic set of economic, social and environmental goals becomes the centre of policy making. Furthermore, this position does not recognise any inherent conflict between capitalism and the achievement of those alternative social and environmental goals as it does not assume that capitalism is dependent on GDP growth (Seidl and Zahrnt 2010; Schneidewind and Zahrnt 2013; van den Bergh 2011; van den Bergh and Kallis 2012; Stiglitz et al. 2011).

This position comprises social-democratic and conservative variants which put different emphasis on the role of the state and social equality. For instance, while Seidl et al. (2010) and Schneidewind and Zahrnt (2013) argue for a green-oriented reform of market capitalism through eco-taxation, sustainable consumption, redistribution of work and green infrastructure investments, Miegel (2011) takes issue with values of greed and excess and advocates the deregulation of the welfare state, restoration of charitable welfare and self-responsibility as solutions.

The second, *anti-capitalist*, position sits at the other end of the spectrum in that it regards capitalism as inherently dependent upon growth. It argues that abandoning growth as economic goal, which it views as necessary to counter environmental destruction and social problems, will inevitably involve a transition away from capitalism (Foster 2011; Latouche 2010; Rätz et al. 2011; van Griethuysen 2010; Bennholdt-Thomsen 2010; Blauwhof 2012). This does not mean an end to the role of markets for the economy, but an abandonment of the profit motive (which feeds, according to this position, the requirement for continuous growth, see Chap. 2), and a greater emphasis on collectively, democratically organised forms of production and the decommodification of labour.

The third position, *alternative-open*, can be seen as a middle way between the other two. On the one hand, it explicitly identifies economic growth as one of the core drivers of environmental and social crisis and therefore advocates degrowth strategies (to reach a sustainable SEE) more clearly than the system-reform position. It also advocates a range of alternative economic and social institutions to establish a sustainable economy, especially through localisation, sufficiency, cooperatives and commons, as well as non-market-based transactions. On the other hand, and this is where it differs from the more radical anti-capitalist stance, this position does not categorically declare capitalism to be the main driver of growth and is hence agnostic as to whether achieving a sustainable state requires a transition away from the current economic system. What is clear, however, is that the alternative institutions and practices that this position advocates would still significantly differ from those under existing forms of capitalism (Paech 2013; Bennholdt-Thomsen 2010; Schneider et al. 2010; Kallis et al. 2012; Dietz and O'Neill 2013; Jackson 2011).

Arguably, a fourth position can be identified which does not argue for postgrowth from a normative perspective but maintains that the combination of a range of conditions and processes will result in a long-lasting, if not permanent phase of stagnant or declining growth rates in Western capitalist countries. For instance, Larry Summers (2016) interprets the phenomenon of "secular stagnation" as a demand-side problem, resulting from an imbalance between excessive savings and insufficient investment (which he thinks can be dealt with through expansionary fiscal policies). In contrast, and as explained in more detail in Chap. 2, Gordon (2012) argues that the capacity for further productivity and economic growth is now much diminished and up against six supply-side "headwinds" of demography, education, inequality, globalisation, energy/environment, overhang of consumer and public debt. While it is difficult to predict how long this phase of stagnant or declining growth rates will last and to what extent it can be counteracted with policies, these diagnoses strongly indicate that Western democracies are well advised to develop alternative institutions to deal with the economic and social repercussions of long-lasting economic stagnation.

Conclusion

This chapter has summarised the main arguments that have featured in debates critical of economic growth. The first that has become accepted even amongst mainstream economists and some political circles is that GDP is not an adequate measure of social and economic progress as it disregards a range of positive contributions to welfare that are not included in monetary terms and excludes many of the costs that emerge from economic growth. More radical strands of growth criticism focus on the detrimental ecological implications of growth and argue that the two cannot be completely decoupled in absolute terms. A range of social issues can also be traced back to growth and the types of societies that growth promotes. More fundamentally, however, it is important to acknowledge that ecological and social implications of growth cannot be separated from each other as ecological destruction is inevitably going to undermine people's wellbeing in the long term. Growth criticism has generated ideas about alternative economic systems which we introduce here under the label of postgrowth: the idea of an SEE and proposals about degrowth. While the chapter demonstrates that this is not a homogenous debate as various, partly conflicting approaches coexist within it, all of these approaches have in common a vision of an alternative socio-economic system which does not prioritise economic growth over other goals and which thus achieves ecological and social sustainability. The next chapter discusses in more detail which visions about the future of wellbeing have been put forward in postgrowth debates and critically examines some of these assumptions.

Note

1. See the latest figures from http://www.worldbank.org/en/topic/poverty/overview.

References

Adler, F., and U. Schachtschneider. 2010. *Green New Deal, Suffizienz oder Ökosozialismus? Konzepte für gesellschaftliche Wege aus der Ökokrise*. München: Oekom.

Anderson, K., and G. Peters. 2016. The Trouble with Negative Emissions. *Science* 354 (6309): 182–183.

Barry, J. 1998. *Rethinking Green Politics: Nature, Virtue and Progress*. Thousand Oaks, CA: Sage.

Ben-Ami, D. 2010. *Ferraris for All. In Defence of Economic Progress*. Bristol: Policy Press.

Bennholdt-Thomsen, V. 2010. *Geld oder Leben. Was uns wirklich reich macht*. München: Oekom.

Black, R., W.N. Adger, N.W. Arnell, S. Dercon, A. Geddes, and D.S.G. Thomas. 2011. The Effect of Environmental Change on Human Migration. *Global Environmental Change-Human and Policy Dimensions* 21: S3–S11.

Blauwhof, F.B. 2012. Overcoming Accumulation: Is a Capitalist Steady-State Economy Possible? *Ecological Economics* 84: 254–261.

Brandt, A.R., Y.C. Sun, S. Bharadwaj, D. Livingston, E. Tan, and D. Gordon. 2015. Energy Return on Investment (EROI) for Forty Global Oilfields Using a Detailed Engineering-Based Model of Oil Production. *Plos One* 10 (12): 18.

Chapman, I. 2014. The End of Peak Oil? Why this Topic is Still Relevant Despite Recent Denials. *Energy Policy* 64: 93–101.

Collier, P. 2007. *The Bottom Billion. Why the Poorest Countries are Failing and What can be Done about it*. Oxford: Oxford University Press.

Costanza, R., J.H. Cumberland, H.E. Daly, R.J.A. Goodland, R.B. Norgaard, I. Kubiszewski, and C. Franco. 2015. *An Introduction to Ecological Economics*. Boca Raton, FL: CRC Press.

Cruz, I., A. Stahel, and M. Max-Neef. 2009. Towards a Systemic Development Approach: Building on the Human-Scale Development Paradigm. *Ecological Economics* 68 (7): 2021–2030.

Czech, B. 2013. *Supply Shock. Economic Growth at the Crossroads and the Steady State Solution*. Gabriola Island: New Society Publishers.
Dale, M., S. Krumdieck, and P. Bodger. 2011. Net Energy Yield from Production of Conventional Oil. *Energy Policy* 39 (11): 7095–7102.
Daly, H., and J.B. Cobb. 1989. *For the Common Good. Redirecting the Economy Toward Community, the Environment, and a Sustainable Future*. Boston: Beacon Press.
Daly, H., and J. Farley. 2011. *Ecological Economics. Principles and Applications*, 3rd ed. Washington: Island Press.
Dietz, R., and D. O'Neill. 2013. *Enough is Enough. Building a Sustainable Economy in a World of Finite Resources*. London: Earthscan/Routledge.
Douthwaite, R. 1999. *The Growth Illusion. How Economic Growth has Enriched the Few, Impoverished the Many and Endangered the Planet*. Totnes: Green Books.
Foster, J.B. 2011. Capitalism and Degrowth—An Impossibility Theorem. *Monthly Review-an Independent Socialist Magazine* 62 (8): 26–33.
Georgescu-Roegen, N. 1971. *The Entropy Law and the Economic Process*. Cambridge, MA: Harvard University Press.
Gordon, R.J. 2012. Is US Economic Growth Over? Faltering Innovation Confronts the Six Headwinds. Working Paper 18315. Cambridge: National Bureau of Economic Research.
Hirsch, F. 1976. *Social Limits to Growth*. Cambridge, MA: Harvard University Press.
IPCC. 2014. *Climate Change 2014: Synthesis Report—Summary for Policy Makers. Contribution of Working Groups I, II and III to the Fifth Assessment Report of the Intergovernmental Panel on Climate Change*. Geneva: Intergovernmental Panel on Climate Change.
Jackson, T. 2011. *Prosperity Without Growth: Economics for a Finite Planet*. London: Earthscan/Routledge.
James, O. 2007. *Affluenza*. London: Vermilion.
Kallis, G., C. Kerschner, and J. Martinez-Alier. 2012. The Economics of Degrowth. *Ecological Economics* 84: 172–180.
Kallis, G., G. D'Alisa, and F. Demaria. 2015. Introduction: Degrowth. In *Degrowth: A Vocabulary for a New Era*, ed. G. Kallis, G. D'Alisa, and F. Demaria, 1–18. London: Routledge.
Kasser, T. 2002. *The High Price of Materialism*. Cambridge, MA: MIT.
Kerschner, C. 2010. Economic De-Growth vs. Steady-State Economy. *Journal of Cleaner Production* 18 (6): 544–551.
Latouche, S. 2010. Degrowth. *Journal of Cleaner Production* 18 (6): 519–522.
Leipert, C. 1986. Social Costs of Economic-Growth. *Journal of Economic Issues* 20 (1): 109–131.

Meadows, D.H. 1972. *Limits to Growth: A Report for the Club of Rome's Project on the Predicament of Mankind*. London: Earth Island.
Meadows, D.H., J. Randers, and D. Meadows. 2004. *Limits to Growth: The 30-Year Update*. London: Earthscan.
Miegel, M. 2011. *Exit. Wohlstand ohne Wachstum*. Berlin: List.
Mill, J.S. and Laughlin, J.L. 1884. *Principles of Political Economy*. New York: D. Appleton & Co.
Murphy, D.J., and C.A.S. Hall. 2011. Energy Return on Investment, Peak Oil, and the End of Economic Growth. In *Ecological Economics Reviews*, ed. R. Costanza, K. Limburg, and I. Kubiszewski, 52–72. Malden: Wiley.
Offer, A. 2006. *The Challenge of Affluence: Self-control and Wellbeing in the United States and Britain since 1950*. Oxford: Oxford University Press.
Paech, N. 2013. *Befreiung vom Überfluss. Auf dem Weg in die Postwachstumsökonomie*. München: Oekom.
Piketty, T., and E. Saez. 2014. Inequality in the Long Run. *Science* 344 (6186): 838–843.
Rahnema, M., and V. Bawtree. 1997. *The Post-Development Reader*. London: Zed Books.
Rätz, W., T. von Egan-Krieger, B. Muraca, A. Passadakis, M. Schmelzer, and A. Vetter. 2011. *Ausgewachsen! Ökologische Gerechtigkeit, Soziale Rechte, Gutes Leben*. Hamburg: VSA.
Rockström, J., W. Steffen, K. Noone, A. Persson, F.S. Chapin, E.F. Lambin, T.M. Lenton, M. Scheffer, C. Folke, H.J. Schellnhuber, B. Nykvist, C.A. de Wit, T. Hughes, S. van der Leeuw, H. Rodhe, S. Sorlin, P.K. Snyder, R. Costanza, U. Svedin, M. Falkenmark, L. Karlberg, R.W. Corell, V.J. Fabry, J. Hansen, B. Walker, D. Liverman, K. Richardson, P. Crutzen, and J.A. Foley. 2009. A Safe Operating Space for Humanity. *Nature* 461 (7263): 472–475.
Rougoor, W., and C. van Marrewijk. 2015. Demography, Growth, and Global Income Inequality. *World Development* 74: 220–232.
Sala-i-Martin, X. 2006. The World Distribution of Income: Falling Poverty and … Convergence, Period. *The Quarterly Journal of Economics* 121 (2): 351–397.
Schmelzer, M. 2015. The Growth Paradigm: History, Hegemony, and the Contested Making of Economic Growthmanship. *Ecological Economics* 118: 262–271.
Schneider, F., G. Kallis, and J. Martinez-Alier. 2010. Crisis or Opportunity? Economic Degrowth for Social Equity and Ecological Sustainability. Introduction to this Special Issue. *Journal of Cleaner Production* 18 (6): 511–518.
Schneidewind, U., and A. Zahrnt. 2013. *Damit gutes Leben einfacher wird. Perspektiven einer Suffizienzpolitik*. München: Oekom.
Schwartzman, D. 2012. A Critique of Degrowth and its Politics. *Capitalism Nature Socialism* 23 (1). 119–125.

Seidl, I., and A. Zahrnt. 2010. *Postwachstumsgesellschaft. Konzepte für die Zukunft*. Marburg: Metropolis.
Stern, N. 2015. Economic Development, Climate and Values: Making Policy. *Proceedings of the Royal Society B-Biological Sciences* 282 (1812): 5–13.
Stiglitz, J.E., A. Sen, and J.-P. Fitoussi. 2011. *Mis-Measuring Our Lives. Why GDP doesn't Add up*. New York: The New Press.
Strauss, B., S. Kulp, and A. Levermann. 2015. *Mapping Choices: Carbon, Climate and Rising Seas—Our Global Legacy*. Princeton, NJ: Climate Central.
Summers, L.H. 2016. The Age of Secular Stagnation What It Is and What to Do About It. *Foreign Affairs* 95 (2): 2–9.
van den Bergh, J. 2009. The GDP Paradox. *Journal of Economic Psychology* 3 (2): 117–135.
van den Bergh, J. 2011. Environment Versus Growth—A Criticism of "Degrowth" and a Plea for "A-Growth". *Ecological Economics* 70 (5): 881–890.
van den Bergh, J., and G. Kallis. 2012. Growth, A-Growth or Degrowth to Stay within Planetary Boundaries? *Journal of Economic Issues* 46 (4): 909–919.
van Griethuysen, P. 2010. Why are We Growth-Addicted? The Hard Way towards Degrowth in the Involutionary Western Development Path. *Journal of Cleaner Production* 18 (6): 590–595.
Wilkinson, R.G., and K.E. Pickett. 2009. *The Spirit Level. Why More Equal Societies Almost Always Do Better*. London: Allen Lane.

CHAPTER 5

Postgrowth and Human Wellbeing

Abstract This chapter reviews the discussion about the relationship between postgrowth and wellbeing. After providing a brief overview of different wellbeing concepts and measurements, it presents and critically discusses the main arguments that have been made in the postgrowth literature regarding the capacity of postgrowth economies to maintain or even improve present levels of wellbeing. This involves a discussion about the concepts of wellbeing that have been applied in the debate so far, as well as a review of the evidence on the relationships between economic growth and contraction on the one hand and subjective and objective wellbeing outcomes on the other. The chapter argues that the concept of basic human needs deserves more attention in this debate as it is compatible with postgrowth frameworks.

Keywords Subjective and objective wellbeing · Human needs Capabilities · Adaptive preferences · Loss aversion

A core component of the postgrowth discourse, which we introduced in the previous chapter, is its position on human wellbeing. In this chapter, we provide an overview of the wellbeing concepts that the postgrowth debate refers to, the arguments that it makes regarding the ways in which wellbeing can be maintained or improved under conditions of post-growth, especially degrowth, and how these claims compare to empirical

M. Büchs and M. Koch, *Postgrowth and Wellbeing*,
DOI 10.1007/978-3-319-59903-8_5

evidence we have so far on the relationships between growth/economic contraction and wellbeing.

A core postgrowth argument is that economic activity should mainly aim at achieving long-term human wellbeing, not economic growth. This is evident from the discussion of the ends of economic activity in ecological economics. For instance, Daly and Farley state that welfare (they use the term interchangeably with "wellbeing") is "the basic purpose of economic activity" (2011: 494), and ecological economist Daniel O'Neill (2012, 2015) adopted Daly's ends-means spectrum such that wellbeing is defined as the ultimate end of the economy.

Before we examine in more detail the concepts of wellbeing that feature in the postgrowth debate and the arguments that are used to support the wellbeing claims mentioned above, we provide a brief overview of some important conceptual distinctions and measurement issues around wellbeing.

Conceptions and Measures of Wellbeing

In the debate on wellbeing, one often finds a distinction between subjective and objective wellbeing. This can relate to either the content of wellbeing (theory) or the assessment of wellbeing (methods), two dimensions that are often confused in the debate (O'Neill 2008: 139–140). Both content and assessment of wellbeing can be subjective or objective, providing us with the following types of wellbeing: hedonic (subjective and objective); evaluative, eudemonic (subjective and objective); and human needs and capabilities approaches (Table 5.1).

Hedonic wellbeing refers to the presence of positive emotions or states of mind such as happiness and the absence of negative emotions or experiences such as sadness, anxiety, anger or pain. This dimension of wellbeing links back to utilitarian philosopher Bentham's idea that the ultimate benchmark for social and economic success should be the "greatest happiness of the greatest number" (Bentham and Harrison 1988). It is usually assessed subjectively, based on people's own assessments of how they feel. Even though rarely used in wellbeing research, emotional states could also be measured more objectively, e.g. based on physical indicators such as heart rate, brainwaves or facial scanning (Davies 2015: 25, 222).

Evaluative wellbeing refers to people's own assessment of their life or specific aspects of their life such as their income, health, relationship

Table 5.1 Concepts and measures of wellbeing

	Subjective wellbeing "content"	*Objective wellbeing "content"*
Subjective assessment	**Subjective hedonic wellbeing** Self-assessment of subjective states/feelings, e.g. happiness, anxiety, etc.	**Evaluative wellbeing/subjective eudemonic wellbeing** Satisfaction with objective wellbeing dimension (general life satisfaction or satisfaction with specific dimensions, such as health, finances)
Objective assessment	**Objective hedonic wellbeing** Physical measures of emotions (brainwaves, heartbeat, sweat, etc.)	**Human needs approaches; capabilities; objective eudemonic wellbeing** Measures of health, education, community engagement, political participation, freedom, social capital, etc.

status and leisure time. The main difference to subjective hedonic approaches is that this perspective puts greater emphasis on cognitive and longer-term evaluations of life dimensions, less so on momentary emotions.

Eudemonic wellbeing refers to the concept of flourishing and fulfilment, of which both subjective and objective assessments are possible. The idea of *eudaimonia* relates back to Aristotle's conception of leading a good life which is centred around the realisation of one's capabilities and thoroughly considered life goals (Ryan et al. 2008: 142). It is from this perspective of capabilities and flourishing that eudemonic wellbeing is considered as an objective account of wellbeing as discussed, for instance, by John O'Neill (2008). Some parallels can be drawn between the universal human needs and capabilities approaches on the one hand and eudemonic approaches to wellbeing on the other. Both perspectives include a similar set of psychological needs which they regard as universal, including for instance autonomy, the ability to relate to others and the need to have some control over one's environment (Doyal and Gough 1991; Nussbaum 2003; Ryan et al. 2008).

However, fulfilment and purpose in life also have important subjective dimensions as they depend on people's perceptions of their life goals and the extent to which they are achieved, which is why eudemonic wellbeing is also sometimes discussed as a type of subjective wellbeing (Dolan and Metcalfe 2012). Ryan, Deci and colleagues have contributed to the discussion about subjective assessments of eudemonic wellbeing with their self-determination theory which assumes—and has shown empirically—that people whose actions are intrinsically motivated, i.e. because they attache inherent value to them, tend to show higher levels of subjective wellbeing compared to those who are extrinsically motivated, for instance, by externally set rewards such as increases of income, status, fame or power (Ryan et al. 2008).

Finally, human needs and capabilities approaches are usually regarded as objective concepts and assessments of wellbeing. Both have criticised subjective wellbeing approaches. The first, widely discussed criticism of subjective assessments of wellbeing highlights that people's preferences and expectations often adapt to circumstances or social norms. For instance, preferences can adapt downwards to limited sets of opportunities such that "objectively" disadvantageous situations are accepted ("to make life bearable in adverse situations" (Sen 1999: 62)). Just think of women who accept and remain in abusive relationships because they think this is "part of women's lot in life" (Nussbaum 2000: 112) or who do not question their limited education opportunities because being more educated does not match with the identities they have been socialised into (ibid.: 62–63, 126–127). Preferences can also adapt upwards in that people may quickly regard raised living standards as the "new normal" and still desire things they do not need, influenced by advertising or social pressures.

The latter point connects to another criticism of hedonic accounts of wellbeing in that they focus on wants, not needs. This is problematic from an environmental perspective because relative wants, in contrast to needs, are insatiable (Gough 2015: 1202; Koch and Buch-Hansen 2016: 31), encouraging an escalation of consumption and resource use/waste.

The insight of adaptive preferences has been extended to a wider critique of societies in which there is considerable pressure to be seen to be "happy", albeit in a superficial way where happiness is linked to consumption, status and other external drivers, fed by a relentless media industry, meanwhile distracting people's attention from deplorable states of injustice and alienation from their own deeper needs (Davies 2015). What is regarded as problematic here, too, is that subjective accounts emphasise individuals' responsibility for their wellbeing as they assume that happiness and satisfaction largely depend on individuals' perceptions and responses to their situation, rather than on wider social contexts. Hence, the state and other collective bodies' responsibility in supporting wellbeing is underplayed (Davies 2015). Subjective assessments of wellbeing can also be criticised from a methodological point of view. They are usually collated through ordinal survey questions which ask respondents to rank their happiness or life satisfaction on a scale (Dolan and Metcalfe 2012). There are several issues with this approach to measuring wellbeing. For instance, it is unclear how comparable subjective responses to such scales are, as people might use different processes of translating their subjective experiences into numbers on a scale and also because norms are likely to differ across cultures regarding the expression of subjective wellbeing (Stearns 2012). Furthermore, using such bounded scales for measurements over time can be limited as they do not leave any "room for improvement" once an individual has already located themselves at the upper end of the scale. This is relevant for discussing the relationship between economic growth and subjective happiness over time below.

In contrast, objective approaches argue that at least certain basic dimensions of wellbeing can be defined independently of individuals' perceptions and cultural context, that people have equal rights to have these basic requirements met, and thus for the state or other collective bodies to share responsibility for generating the conditions for this to happen. Two main lines of thinking can be identified here, the capabilities approach developed by Amartya Sen (e.g. 1999), Martha Nussbaum (2000, 2003) and others; and the universal human needs approach, developed by Max-Neef (1991) and Doyal and Gough (1991).

The capabilities approach, which was initiated by Amartya Sen, explicitly criticises purely subjective approaches to human wellbeing. For instance, it points out that people often adapt preferences to their situation, including those where freedoms and capabilities are limited. It is therefore possible for someone to be subjectively satisfied with their situation despite being poor, exploited or otherwise limited in their freedom. Moreover, certain types of preferences are not just the result of adaptations to disadvantageous situations but of deeper lying processes of internalisation and socialisation (e.g. when it comes to gender roles) (Nussbaum 2011: 83–84).

Capabilities can be defined as the effective opportunities that are available to people to achieve a range of "functionings". Scholars in this field differ regarding their position to how "objective" or "subjective" they define capabilities and functionings to be. For instance, Sen has taken a more liberal stance as he defines functionings as "the actions and activities that they [people] want to engage in, and be whom they want to be" (Robeyns 2005: 95), leaving room for subjective determination. Nussbaum criticised Sen's approach, arguing "Sen's 'perspective of freedom' is too vague. Some freedoms limit others; some freedoms are important, some trivial, some good, and some positively bad" (2003: 33). Rather, she argues, "commitments about substance" (ibid.) need to be formulated, and hence endorses a list of basic capabilities that people have a right to achieve (Nussbaum 2000, 2003). In her view, certain capabilities are so fundamental that they need to be achieved for someone to be able to live a life of "dignity", including the ability to live life to its normal length; bodily health; bodily integrity; to use one's senses, imagination and thought in a "humanely" and educated way; to be able to experience the full range of emotions; practical reason and the ability to use it critically; the ability to affiliate oneself with other human beings; the ability to live "with concern for and in relation with" other species; the ability to play; and the ability to have control over one's environment, both politically and in terms of property rights (Nussbaum 2003: 41–42).

The theory of universal human needs developed by Doyal and Gough (1991) builds on Max-Neef's et al. (1991) distinction between needs and needs satisfiers—where the former are regarded as universal and the latter as culturally and historically variable. Max-Neef had developed a matrix of interrelated and non-hierarchical needs and needs satisfiers consisting of nine dimensions of "axiological needs"—subsistence, protection, affection, understanding, participation, identity, idleness, creation and freedom—and four dimensions of "existential needs"—being, having, doing and interacting. In contrast, Doyal and Gough (1991) argued for a hierarchical set of needs, reaching from the universal goal of "minimally impaired participation in society", to basic needs including physical health and "autonomy of agency", to universal types of needs satisfiers, including amongst others "adequate nutritional food and water", housing, physical security and education, through to universal social preconditions for the satisfaction of these needs (Doyal and Gough 1991: 170). While this approach acknowledges that needs satisfiers are culturally specific and therefore flexible, it regards this set of needs as universal—valid across time and space. At its core is a criticism of subjectivist and relativist approaches to needs, which they see as often being exploited by neoliberal strategists who argue that needs satisfaction can be left to the market. Instead, they argue that the satisfaction of universal needs is a human right for which the state should take at least some responsibility (Doyal and Gough 1991). In a recent paper, Gough (2015) has discussed the relationship between the universal human needs approach and climate change. Here, he emphasises that human needs are not only universal across different cultures, but also across generations and that therefore current generations are morally obliged not only to fairly distribute resources amongst the current generation, but also to "hand down" conditions that enable the satisfaction of future generations' needs. With reference to the context of climate change, he deduces from this the obligation to set a planetary greenhouse gas emissions ceiling and reduce emissions accordingly, to fairly allocate emission quotas amongst countries and people, and to fund adaptation and compensation measures for those already affected by climate change (Gough 2015).

This overview demonstrates that much internal debate remains about the conceptualisation of objective wellbeing. In addition, the "objective" measurement of objective wellbeing remains a dynamic field of discussion and research. As already indicated above, there has been a lot of work to develop indicators of objective wellbeing, even though some of those indicators can only ever crudely capture some of the dimensions of universal needs or capabilities discussed above. In ecological economics, a lot of research has also been undertaken to develop indicators and indices that can function as alternatives to GDP to measure human wellbeing at the national level. Very broadly, three different approaches can be distinguished here (but note that some of them also include subjective assessments of wellbeing as part of a larger set of indicators) (Fleurbaey and Blanchet 2013). *First*, indices that combine different dimensions of wellbeing but make them comparable by translating them into monetary values—often subtracting value to account for costs such as environmental damage. Since they express human welfare in monetary terms, these indices enable comparison to developments of GDP. A prominent example is the Index of Sustainable Economic Welfare (Daly and Cobb 1989) which was later revised and renamed as Genuine Progress Indicator (GPI) (Talberth et al. 2007). The GPI is based on personal consumption expenditure which is closely related to GDP but adjusted through a range of other measures including the cost of environmental damage and crime, the positive benefits from activities in the informal economy, and income inequality. *Second*, indicators that combine several indicators into an index without converting them into monetary values or even harmonising the different dimensions. Examples for this approach are the Human Development Index which is based on the capabilities approach and includes GDP per capita, life expectancy and education; or the Index of Economic Well-Being which combines per capita consumption, measures of stocks of various types of capital, including environmental and human capital, measures of income distribution and of economic insecurity from unemployment, family breakup or ill health (Osberg and Sharpe 2002: 295). *Third*, there are indices that combine objective and subjective measures of wellbeing such as the Happy Planet Index developed by the New Economics Foundation which mixes "happy life expectancy" (life expectancy multiplied by a measure of happiness) and the ecological footprint (NEF 2016); and the OECD's Better Life Index which is an interactive tool to assess wellbeing across OECD countries using indicators for 11 wellbeing domains.[1]

As we will demonstrate in more detail below, the postgrowth literature draws on a variety of wellbeing concepts. We argue here that the debate would benefit from more coherent conceptual positioning, in particular, by drawing more extensively on the basic human needs approach while also considering interrelations between objective and subjective (assessments of) wellbeing. As other scholars have started to highlight (Gough 2015; Koch et al. 2017; O'Neill 2011), the concept of basic human needs is very relevant for postgrowth research on wellbeing as it fits with ideas of non-substitutability and satiability which are part of a strong sustainability framework. Non-substitutability means that needs satisfaction consists of various dimensions which cannot be substituted (for instance, a lack of supportive and fulfilling relationships cannot be replaced by rising income or consumption). The concept of satiability states that needs satisfaction can be achieved within certain limits of material throughput because from certain points onwards, further increases in income or consumption do not further improve needs satisfaction. In addition, the concept of satiability can be connected to specifying limits of material throughput such that the satisfaction of future generations' needs is provided for.

We believe that this is a key component of human needs theory in the context of postgrowth which can draw on work by several scholars who have connected discussions on needs and capabilities with thinking about intergenerational justice and climate change impacts. For instance, authors such as Page (2007) have extended Nussbaum's list of central capabilities to "include the capability to experience life in an environment devoid of dangerous environmental impacts such as those associated with climate change" (ibid.: 464). If human needs are universal, they also apply to future generations. Therefore, this additional capability would act as a limitation on other central capabilities or needs as it establishes the rule that the fulfilment of current generations' needs to preserve the capability of future generations to live in a life-sustaining environment.

The idea that the fulfilment of needs in the present is constrained by the requirement to preserve the conditions for wellbeing for future generations is also directly supported by the framework of universal human needs. In his recent paper, Gough (2015) argues that this approach implies a moral perspective which grants rights to the fulfilment of needs not only to current but also to future generations. Doyal and Gough's *Theory of Human Need* (1991) expands this argument by explaining that "a commitment to a contemporary moral vision of the good makes little sense applied only to present generations. To so damage the environment as to jeopardise the long-term survival of a form of life which we believe embodies the good is to renounce our commitment to that good—no more and no less" (ibid.: 145). In other words, they argue here that the "vision of the good"—which is at the core of the idea of avoidance of serious harm and hence of the ultimate wellbeing goal in this framework—actually needs to embrace consideration of the wellbeing of future generations (as well as of other cultures/countries in the same time frame) to be a valid concept in the first place. As Koch and Buch-Hansen (2016: 35) point out, Gough has also accepted the critique of the framework's earlier endorsement of levels of needs satisfaction based on the highest standards achieved in the Western world—here Sweden—and conceded that needs satisfaction may need to be constrained to "less than optimal generalizable levels". This is also supported by thinkers such as Clark Wolf who states: "Where our present activities are not necessary for satisfaction of present fundamental needs, and put at risk the basic needs of future generations, then they are unjust" (2009: 373). Philosopher Baxter (1999) even goes as far as to argue that principles of intergenerational justice imply a moral duty for current generations to *reduce* their material living standards to provide for basic needs satisfaction of future generations.

An open question remains, however, where exactly the premise of limiting the level of needs satisfaction to a level of material throughput that enables needs satisfaction of future generations should be linked to the existing framework of basic human needs. We argue here that the best place for this linkage is at the highest level of the framework which defines the ultimate goal of wellbeing as the avoidance of serious harm. More precisely, serious harm is defined as the "fundamental disablement in the pursuit of one's vision of the good, whatever that vision is" (Gough 2015: 1196). Since in theory, that vision could be defined such that it requires maximum levels of needs satisfaction with resource throughputs at unsustainable levels, this is the level at which resource limitations for the sake of future generations' wellbeing need to be considered. This re-framing of ultimate wellbeing goals will then also help to determine the available level of material throughput for the satisfaction of all other intermediary needs in concrete, democratic decision-making.

Even though we fully endorse a focus on the framework of basic human needs in this debate—which is understood primarily as an "objective" account of wellbeing—we think the postgrowth and wellbeing debate would also benefit from greater clarity about the relationships between objective and subjective wellbeing. As the discussion above has shown, the distinction between objective and subjective wellbeing is often not as clear-cut as commonly presented. First of all, subjective and objective wellbeing are likely to be closely and bi-directionally interlinked. For instance, a large body of research is concerned with the determinants of subjective wellbeing. It shows that several objective dimensions of wellbeing, especially health and supportive relationships (Deeming 2013; Helliwell et al. 2015), positively contribute to people's happiness and life satisfaction. However, the relationship can also work the other way round, for instance, Argyle (1997) found that high subjective wellbeing supports people's health (Argyle 1997), and it is plausible to assume that it also affects other dimensions of objective wellbeing, including people's relationships, employment status, education and income because "happier" people might be more motivated, easier to interact with, etc.

A second point relates to the ways in which wellbeing is generated through social practices (on which more in Chap. 6) and the acknowledgement that it is both individuals with their subjective perceptions, as well as wider social contexts which co-produce individual and collective wellbeing. From this follows that responsibility for wellbeing cannot be wholly individualised or collectivised, but remains to be shared. Furthermore, a social practice-based understanding of the generation of wellbeing also highlights that the transition to a postgrowth society inevitably needs to involve a collective redefinition of the main aims of wellbeing—for instance along the lines of the fulfilment of basic human needs and eudemonic wellbeing. Without such a collective redefinition and acceptance of wellbeing aims, people are likely to respond negatively to restrictions to GDP growth and related contraction of consumption opportunities.[2]

Wellbeing in the Postgrowth Discourse

The predominant view put forward by the postgrowth literature is that human wellbeing can be maintained, or even improved, in the context of an SEE and even the phase of degrowth. One example for this view is Schneider's (2010: 511) often-cited definition of degrowth as an "equitable downscaling of production and consumption that increases human well-being and enhances ecological conditions at the local and global level, in the short and long term". This definition draws on early accounts of the idea of degrowth, for instance by André Gorz who stated that it is now a sign of "realism" to "advocat[e] greater wellbeing through the inversion of growth and the subversion of the prevailing way of life" (Gorz 1980: 14).

This section first reviews the arguments that have been put forward to support the assumption that wellbeing can be maintained or even improved under zero- or degrowth. For a radical social movement, it is strategically important to emphasise expected benefits from advocated changes and to formulate a utopian vision of the future to gain popular support and political traction. However, and this is what we will argue here, it is equally important to be realistic about potential wellbeing implications of postgrowth, so that we can develop alternative institutions that can tackle possible negative implications and establish a transformed social system. We will therefore also review alternative evidence, especially on wellbeing implications of economic contraction, to further examine these arguments. Other degrowth proponents may argue that this is not a legitimate way of examining these issues because degrowth is meant to be a voluntary and democratically designed process, underpinned by new institutions, rather than an involuntary economic crisis within the existing setting. This is a valid point, however, since we cannot anticipate future institutional structures, functioning of the economy and social responses, looking at historical data on relationships between change of GDP and wellbeing outcomes is the best available source of information that currently exists to alert us of potential problems.

Most fundamentally, the argument that postgrowth will benefit wellbeing is based on a different time perspective compared to that usually applied in public debates and economic reasoning. Essentially, postgrowth proponents adopt a long-term perspective which maintains that if our economies keep growing, this will eventually undermine the ecological and physical basis of our existence as climate change impacts increase, resources become exhausted and vital ecosystems become dysfunctional. This is evident from the first degrowth conference declaration: "There is an eventual limit to the scale of global production and consumption and to the scale national economies can attain without imposing environmental and social costs on others elsewhere or future generations. (…) If we do not respond to this situation [of exceeding ecological limits] by bringing global economic activity into line with the capacity of our ecosystems (…), the result will be a process of involuntary and uncontrolled economic decline or collapse" (Research and Degrowth 2010: 523). It is implied in this statement that "uncontrolled economic decline or collapse" would result in considerable deterioration of human wellbeing.

The second core argument in the debate is that—beyond a certain level of living standards—wellbeing often stagnates or even decreases with GDP growth. Many authors in the postgrowth debate (Schneider et al. 2010: 512; Alexander 2012: 354) refer here to notions of subjective wellbeing and what has become known as the Easterlin paradox (Easterlin 1974). The Easterlin paradox showed, initially for the US, that while higher income was related to higher levels of happiness in cross-sectional analysis, aggregate happiness scores for the US did not increase over time despite rising levels of GDP. This paradox has since been confirmed in numerous studies for other countries (Easterlin et al. 2010; Blanchflower and Oswald 2004; Diener and Seligman 2004: 5–6; Layard 2005: 30) and, according to a study by Lane (2000), subjective wellbeing in the US even declined over time despite rising levels of GDP. However, some issues need to be considered here as cross-sectional analysis still confirms a relationship between GDP and wellbeing (Fritz and Koch 2016; O'Neill 2015), especially if GDP is measured on a log, not a linear scale (Deaton 2008), which may be more appropriate since GDP often increases exponentially, while wellbeing is measured on a bounded scale. The fact that subjective wellbeing is measured with bounded scales could even be an important reason for its seemingly widening distance from GDP over time.

These concerns have found little attention by growth critics who explain the apparent phenomenon of widening gaps between subjective wellbeing and GDP over time in various ways. The first explanation refers to "hedonic adaptation" or the "hedonic treadmill": the idea that people quickly adjust their expectations to raised standards of living—gains are quickly taken for granted (Brickman and Campbell 1971). For instance, research on people winning the lottery has shown that any changes in subjective wellbeing following the event are short-lived (Brickman et al. 1978). Another reason given for findings that support the Easterlin paradox is that, once basic needs have been satisfied, consumption is more about status competition rather than the satisfaction of needs. Therefore, in richer societies, relative income—how much more or less one earns than other people in society—and what one is therefore able to consume compared to them—becomes more important than absolute levels of income (Layard et al. 2010). And if everyone's incomes increase by a similar proportion, relative income does not change.

The third type of explanation for stagnating or falling subjective wellbeing in rich countries despite rising levels of GDP relates to the role of values. On the one hand, research by Inglehart (1981) suggests that post-materialistic values become more important in rich societies: if people do not prioritise income and living standards in their conceptions of life goals, rising levels of income will not contribute to increasing subjective wellbeing. However, if people have adopted post-materialistic values, stagnating or even falling subjective wellbeing in rich countries over the last few decades would then also indicate that these alternative values are not being fulfilled either. On the other hand, scholars such as Alexander (2012: 350) assume that materialistic values still prevail in Western societies and argue that materialism is related to lower levels of subjective wellbeing as also highlighted by various other studies (Kasser and Ryan 1993, 2002).

The fourth explanation for stagnating or declining subjective wellbeing in rich countries puts greater emphasis on certain social processes or characteristics that growth-based societies seem to nurture. This includes an acceleration of the pace of life, increasing levels of competition and hence stress and potentially even burnout, which some authors see as a postmodern form of alienation (Rosa and Trejo-Mathys 2013). Other authors see the main reason for declining levels of subjective wellbeing in the deterioration of personal relationships or "companionship" (Lane 2000). This can partly be explained through an increasing individualisation and marketisation of society which reduce the importance of close family, friendship, neighbourhood or community relationships as many services can now be purchased on the market, with the result that people spend more time in marketised interactions.

Part of the postgrowth debate about stagnating or declining living standards despite GDP growth also focuses on objective measures of welfare, such as the Index of Sustainable Economic Welfare (ISEW) and the updated Genuine Progress Indicator (GPI). For instance, during the late 1980s Max-Neef (1995) developed his "threshold hypothesis"—that wellbeing in rich countries declines once a certain income threshold is reached. His analysis showed that while the ISEW developed by Daly and Cobb (1989) first rises with increasing GDP, it declined from a certain point onwards despite continuing GDP growth in the US, UK, Germany, Austria and the Netherlands. A multitude of studies followed, using either the ISEW or the GPI, showing the same patterns for a variety of countries or regions, including Europe (Jackson and McBride 2005): 17 countries across the world (Kubiszewski et al. 2013), France (Fleurbaey and Blanchet 2013: 10), the growth-latecomer Asia-Pacific region where the threshold occurred at lower levels of GDP (Lawn and Clarke 2010) and other previous studies in 21 different countries or regions and two US states (Posner and Costanza 2011). These findings are explained with the rising social and environmental costs that GDP growth generates, as the ISEW and GPI subtract these costs from GDP.

These results are very relevant in this context, but arguably still focus too much on an economic notion of wellbeing. An alternative simple, "objective" measure of wellbeing that is also of interest here is life expectancy—an easily measurable indicator of population health. While research on the relationship between economic growth and life expectancy has generated conflicting results, there is some evidence that suggests that, beyond a certain level of development, growth contributes little to increases in life expectancy. Generally speaking, when it comes to the relationship between the level of income and life expectancy, cross-sectional studies show a very similar pattern to that for subjective wellbeing discussed above: while life expectancy and GDP are generally positively correlated (Preston 1975; Easterly 1999; Fritz and Koch 2014, 2016; Pritchett and Viarengo 2010), this relationship tends to be weaker amongst countries with high levels of GDP if income is measured on a linear (rather than log) scale (Wilkinson and Pickett 2009; Kangas 2010). This suggests that life expectancy gains flatten out beyond a certain level of development, and some research has even demonstrated that relatively high levels of life expectancy can be achieved at "moderate" levels of energy use and emissions (Steinberger and Roberts 2010; Lamb et al. 2014). However, it is unclear whether this "levelling off" phenomenon occurs because of decreasing marginal impact of GDP on improving life expectancy or because there are "natural" limits to the expansion of the human life span as some scholars have suggested (Dong et al. 2016).

Findings are more inconsistent when it comes to the impact of GDP growth on life expectancy in early phases of economic development, but they suggest that social and institutional context matters for this relationship. For instance, one hypothesis is that in early phases of development, economic growth enables countries to invest more in infrastructures that improve hygiene, in health services and education, all of which contribute positively to improving life expectancy. This hypothesis is empirically supported by various studies (Jamison et al. 2013; Hertzman and Siddiqi 2000; Granados 2012). However, other historical research has shown that there have also been periods of rapid GDP growth in earlier phases of development which had negative impacts on life expectancy, for instance in Sweden (Granados and Ionides 2008) and England/Wales (Granados 2012) in the first half of the twentieth century. Another study on England argued that rapid growth led to negative health impacts during industrialisation due to a breakdown of urban administration and health services, evident from rising deaths from diseases related to insufficient sanitation and overcrowding, and that the breakdown of services could be explained by a disruption of "established social relations, ideologies and structures of authority" (Szreter 1999: 148). This would support the assumption that economic growth only improves life expectancy if the context enables investments into infrastructures and institutions that improve population health.

The third core argument put forward in the postgrowth debate to support the vision that human wellbeing can be maintained or improved under degrowth focuses on the redefinition of wellbeing. This perspective criticises economic measures of wellbeing, as well as those that are based on happiness as too narrow and inappropriate in the face of climate change and limits to economic growth. Instead, this perspective advocates broader understandings of wellbeing which largely relate to concepts of eudemonic wellbeing (Ryan et al. 2008), human flourishing and capabilities (Sen 1999; Nussbaum 2003). For instance, references are made in the postgrowth and growth-critical literature to ideas of "joy of living" (Schneider et al. 2010: 513), the "good life" (Muraca 2012), "voluntary simplicity" (Elgin 1982) or "alternative hedonism" (Soper 2008) as well as alternative conceptions of prosperity which do not depend on economic growth and consumption (Cassiers 2015; Jackson 2011). They all have in common the idea that a fulfilling and prosperous life does not depend on high income and consumption or other external markers of success but on meaning and purpose in life, the opportunity to become the kind of person one aspires to be, on supportive relationships, etc.

Connected to this perspective is an emerging debate (e.g. Koch and Buch-Hansen 2016) in the growth-critical literature that the ultimate goal of economic activity should be the fulfilment of basic human needs, possibly below the level of "optimum" needs satisfaction as initially suggested by Doyal and Gough (1991). In fact, this idea connects back to the original definition of the aims of degrowth as formulated by the first degrowth conference in Paris in 2008 which stated: "The objectives of degrowth are to meet basic human needs and ensure a high quality of life" (Research and Degrowth 2010: 524). In the theories developed by Doyal and Gough and Nussbaum, basic human needs include some of the dimensions that connect to ideas of human flourishing, the good life and capabilities, for instance the ultimate goal of autonomy and full participation in society (Doyal and Gough 1991), or Nussbaum's (2003: 41–42) central capabilities of "senses, imagination and thought", "emotions", "practical reason" and "affiliation".

The postgrowth discourse not only applies these ideas of alternative concepts of wellbeing to wealthy Western societies, but also to developing countries. The argument here is that defining development as economic growth equates to a Western "colonisation" of the idea of development. While of course hunger, malnutrition, poverty, etc., need to be overcome, proponents of post-development (Rahnema and Bawtree 1997) or "reflexive development" (Pieterse 1998) argue that postgrowth is also about creating new conceptual spaces for people in developing countries to define development from their own perspective.

In a nutshell, therefore, those who advocate a redefinition of wellbeing argue that the context of postgrowth will provide much better conditions for achieving the aims defined by alternative conceptions of wellbeing—flourishing, the good life, etc.—because it will create societies that are less commodified and marketised, enabling people to pursue more meaningful activities and have more time to interact in supportive relationships, engage in the community and public life, etc., and societies that are less competitive and more equal, thus reducing stress and status anxiety which can positively contribute to mental and physical health (Jackson 2011: 16; Paech 2013; Schneider et al. 2010).

Critical Review of Postgrowth Arguments on Wellbeing

All of these arguments that support the vision that wellbeing can be maintained or improved under postgrowth are valid as, essentially, they point to the ways in which environmental and social costs of growth-based societies already affect human wellbeing in current societies and might undermine it more fundamentally in the long term. However, it is also important to scrutinise them critically—with the aim to strengthen them and further ideas about the kinds of institutions and social conditions that can support wellbeing under postgrowth.

One way of doing this is to evaluate available evidence on the impacts of economic crises on people's subjective and objective wellbeing. Of course no direct conclusions can be drawn from this regarding the ways in which postgrowth might impact on wellbeing because, by definition, postgrowth is not equivalent to economic crisis but a voluntary and democratically negotiated transition of society and economy—the "prosperous way down" (Odum and Odum 2006). However, we argue that comparing outcomes across different contexts in the past can provide useful insights into ways in which institutions can mediate wellbeing implications of contracting economies.

The first argument that needs to be considered here relates to one of the explanations for the phenomenon that people's subjective wellbeing does not increase over time despite continuing economic growth. The idea of hedonic adaptation assumes that people's expectations quickly adjust upwards to take higher living standards for granted. The question is whether the same adaptation occurs when living standards *decrease*. This is an implicit assumption in the postgrowth literature, supported by writings on adaptive preferences which highlight the adjustment of expectations to adverse circumstances (Sen 1999: 62; Nussbaum 2000: 112). Recent empirical research found that values of subsistence and security increased (interpreted as a downgrading of life goals), especially amongst economically deprived groups, following the recession in the late 2000s (Austin 2016).

These are important findings and assumptions. However, the question is how much they can tell us about people's wellbeing responses to situations of (relatively rapid) decreases of material living standards. The theory of adaptive preferences tends to focus on groups who find themselves in long-term situations of limited opportunities, not on situations in which opportunities (rapidly) deteriorate, and it is unclear from Austin's research whether these values shifts had any consequences for people's wellbeing.

In this context, it is important to take the concept of "loss aversion" (Tversky and Kahneman 1991) into consideration—the idea, developed based on empirical research, that people respond more strongly, and negatively, to losses than to gains. One early example was research on hedonic adaptation which found that lottery winners were not happier than people in a control group. In contrast, people who had become paralysed in an accident were significantly less happy compared to a control group (Brickman et al. 1978). The authors explain this finding with hedonic adaptation to a new normal amongst lottery winners, compared to a "nostalgically" positive view of the past amongst accident victims. If the theory of loss aversion is right, processes which imply losses of (consumption) opportunities are likely to have negative impacts on people's wellbeing compared to processes that offer gains.

A lot of research on the impact of economic crises on people's subjective and objective wellbeing confirms loss aversion theory. For instance, recent research demonstrates that recessions can affect subjective wellbeing negatively, at least in the short term. For instance, subjective wellbeing decreased in Greece, Syria and Egypt (Diener and Tay 2015: 139), in the UK and Germany (Mertens and Beblo 2016) and in transition countries of Eastern Europe, Caucasus and Central Asia (Habibov and Afandi 2015) following the 2008 economic crisis. Similarly, Fanning (2016: 100) who correlated change of growth and change of life satisfaction scores for 116 countries between 2005 and 2015 found that while there was no association between the two measures in countries with positive growth rates (consistent with hedonic adaptation), there was a significantly negative association for countries with negative growth (recessions), which is consistent with loss aversion theory.

Similar results emerge if one focuses on objective measures of wellbeing as there is evidence that recessions tend to have negative impacts

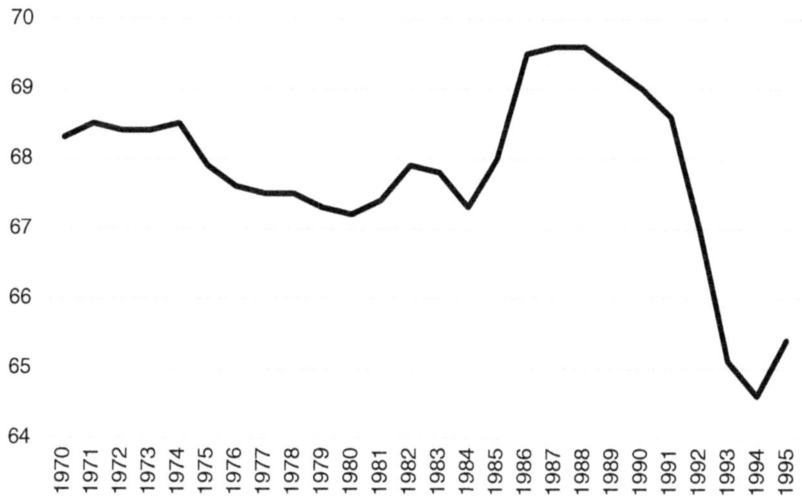

Fig. 5.1 Life expectancy (at birth, whole population) in the Russian Federation. *Source* OECD Health Statistics

on people's health and life expectancy. Research has shown that people who suffer job loss and a related decrease in income are generally in poorer health compared to their counterparts (McKee-Ryan et al. 2005; Catalano and Bellows 2005). Moreover, their life expectancy can be affected for a long time following a recession (Sullivan and von Wachter 2009). Health-harming behaviours, and certain illnesses and causes of death seem to increase during recessions. For instance, studies found that often during recessions mental health deteriorates (Bacigalupe et al. 2016; Zivin et al. 2011) while smoking and drinking as well as suicides increase (Suhrcke and Stuckler 2012; Chang et al. 2013, Haw et al. 2015). This can translate into decreasing life expectancy, for instance in Russia (Gavrilova et al. 2000) and Central and Eastern Europe following the economic recessions that these regions experienced triggered by the collapse of communist regimes in the early 1990s (Hertzman and Siddiqi 2000). The example of Russia is particularly extreme as it experienced a drop in life expectancy of more than 5 years in the early 1990s (see Fig. 5.1) (Parsons 2014: 2). This was unprecedented for a country not at war and is only comparable to one other example in modern times:

sub-Saharan Africa which suffered a similar decrease of life expectancy due to the HIV/AIDS epidemic (ibid.).

The impact of a contraction of national income on health could also be related to how rapidly this is happening, e.g. some scholars have suggested that health effects of recessions are more negative if they are more rapid (Karanikolos et al. 2013: 1326; Bohk and Rau 2015).

However, it needs to be stressed that evidence is not clear-cut as in some cases life expectancy improved following a recession (Gerdtham and Ruhm 2006; Ruhm 2000; Stevens et al. 2015) or was not affected to nearly the same extent than seen in Russia or other Eastern European countries in the early 1990s. For instance, in Cuba, which experienced a severe economic crisis following the collapse of the Soviet Union, life expectancy remained relatively stable (Borowy 2013). How can these contrasting findings be explained? One possibility is that the increase of health-harming behaviours during or after a recession primarily occurs in smaller groups of people who are worse off financially, unemployed or at risk of job loss. In contrast, the majority of the population who remains in employment may take up healthier behaviours as they find more time for sleep and exercise due to reduced working hours or cut down smoking and alcohol intake if governments increase taxes on these substances during a recession. Reduced working time also limits exposure to hazardous work environments, and a decline of traffic decreases the risk of road accidents (Karanikolos et al. 2013: 1325; Stevens et al. 2015; Suhrcke and Stuckler 2012). Another explanation is that spending on health services is often counter-cyclical which could link to higher death rates during times of growth, especially amongst the elderly and residents in nursing homes (Stevens et al. 2015 for 1972–2006 in the US). A well-resourced health system and cooperative social institutions as well as low social inequality have been identified as important factors to explain the situation in Cuba discussed above (Borowy 2013). Generally, the level of welfare spending has been found to increase life expectancy at given levels of GDP (Kangas 2010; McGuire 2001; Beckfield et al. 2015) and to reduce negative impacts on life expectancy during economic crisis (Bacigalupe et al. 2016; Karanikolos et al. 2013; De Vogli and Owusu 2015).

This overview of conflicting evidence on the relationship between health outcomes and economic contraction suggests that it will be important to identify the contextual factors that mediate those outcomes, including levels of employment, inequality and social spending.

When it comes to unemployment, the usual pattern is that it increases during times of recession because firms reduce production and hence lay off workers to save costs. However, some economic models predict it might be possible to keep relatively high employment levels during economic stagnation by cutting working time and redistributing work (Victor 2012; Victor and Rosenbluth 2007).

Another concern relates to the second core argument made in the postgrowth literature: that postgrowth societies will open up new opportunities to improve people's wellbeing if alternative concepts of wellbeing are adopted. The concern here is that this will only work if people collectively develop and adopt such alternative conceptions of wellbeing and adjust their expectations accordingly—so that effects of loss aversion are avoided. This also relates to the observation that people who currently live slower and more sufficient lives, e.g. without cars, no or fewer flights, reduced mobile phone and computer use, etc., are exposed to exclusionary dynamics in high-speed, growth- and acceleration-oriented societies (Muraca and von Egan-Krieger 2011: 51). The pressure for academics to regularly fly to attend international conferences can serve as a concrete example. International conference attendance helps academics to build international networks which are crucial for being in the position to develop multinational collaborations and grant applications. Being "internationally recognised" as a scholar—measured by exactly these types of activities—is often a criterion for promotions or job applications. Therefore, academics, especially more junior ones, put themselves at a considerable disadvantage if they refrain from flying to international conferences in an effort to reduce their carbon footprint.[3] And being socially disadvantaged is likely to have negative impacts on people's wellbeing. It therefore seems essential that deceleration and degrowth-related lifestyle changes are distributed evenly in society to avoid such exclusionary processes. Possibly, living a more fulfilled, slower life will only contribute to collectively stable or even rising wellbeing if people perceive everyone else to "sit in the same boat".

This concern links to the final point that needs to be discussed in relation to the prospect for maintaining or improving wellbeing under postgrowth: an important assumption in postgrowth debates is that greater levels of equality will support wellbeing outcomes in phases of economic contraction (see Chap. 7 for further details). Several scholars have shown that levels of inequality in organisations or society have a detrimental effect on people's health and life expectancy (Wilkinson and Pickett

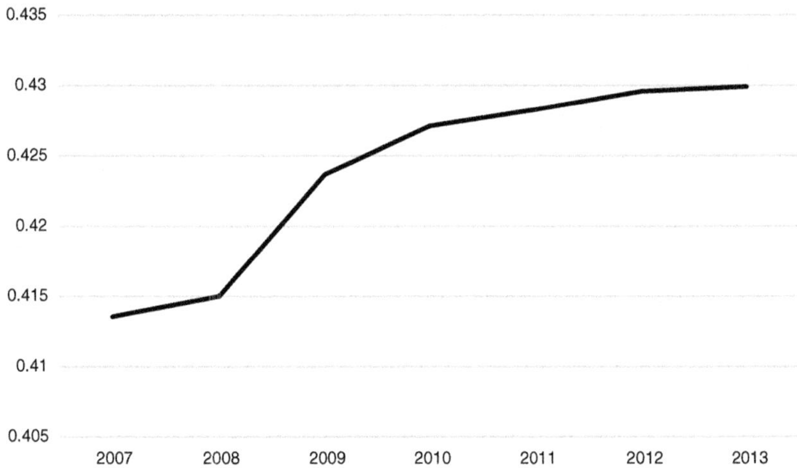

Fig. 5.2 Income inequality in the OECD since the 2007/2008 financial crisis. *Source* OECD Inequality update 2016 http://www.oecd.org/social/inequality-and-poverty.htm

2009; Marmot et al. 1991; Rosero-Bixby and Dow 2016). A common explanation for this phenomenon is that inequality increases the levels of stress that people experience, in both lower and higher social positions, as it increases competitive pressure and anxiety about status maintenance. Reducing levels of inequality is therefore likely to be an important precondition for achieving more favourable wellbeing outcomes during postgrowth. This is especially so since inequality has risen in various countries over the last decades. For instance, the share of wages in GDP has gradually fallen since the early 1960s in the UK (Heatley 2014: 20), and inequality measured by the Gini coefficient[4] has gradually increased or at least remained level (ibid 21–22)—and risen again since the 2007/2008 financial crisis (Fig. 5.2). Recent historical research has also shown that income and wealth inequality in Europe and especially in the US have now reverted back to high levels of the early twentieth century, after it had decreased during the first half of the century and especially since the Second World War (Piketty and Saez 2014).).

Piketty and Saez' (ibid.) explanation for this phenomenon is that economic growth rates have declined since the 1970s, while the rate of return on capital (profits) has declined less so. If this interpretation is correct, it means that economic growth can keep inequality in check and that economic contraction is bad news for inequality. Even supporters of the postgrowth debate have stated that "no growth or low growth has a tendency to deepen social and economic inequalities" (Blewitt and Cunningham 2014: x).

However, it is unclear to what extent growth is required to keep inequality levels in check. Based on modelling work, Jackson and Victor (2016) have recently argued that inequality can be limited even without growth as the impact of growth on inequality also depends on the elasticity of substitution of capital for labour (i.e. technological development), as well as different savings rates of "capitalists" and "workers" (and other types of inequalities like asset ownership). Since these dimensions are also mediated by policies, the postgrowth literature generally assumes that more redistributive policies can keep inequality in check. However, there is disagreement how they are going to be financed without growth.

Additional "softer" social factors are also likely to matter for wellbeing outcomes during times of economic contraction. For instance, one study found that higher levels of institutional and interpersonal trust can reduce declines of life satisfaction during economic crisis (Habibov and Afandi 2015). This is also supported by research which argues that wellbeing is positively influenced by good social relations/social capital (Diener and Seligman 2004).

Conclusion

Postgrowth visions promise that it will be possible to maintain or even improve human wellbeing without growth. First and foremost, this is based on the insight that, in the long term, the environmental and social costs associated with growth will undermine the conditions for human wellbeing. It is also based on evidence that shows that in rich countries, subjective and objective wellbeing do not increase or even decrease with growing GDP over time (while evidence from cross-sectional analysis is less clear). Postgrowth discourses also propose radical redefinitions of wellbeing. We have argued in this chapter that while these are important insights, it remains unclear how economic contraction will impact upon wellbeing as people might respond more negatively to decreasing living standards due to loss aversion. What seems to be clear, however, is that a transition to a postgrowth society will require far-reaching changes of institutional frameworks, including more redistributive policies, and collective conceptions of wellbeing (which includes people's aspirations, values and identities) to counter-act possible negative wellbeing implications of contracting economies. From this perspective, we think it is important to understand more generally the possible ways in which social change can impact on people's wellbeing. The next chapter discusses which insights social practice theory can provide us with for this question.

Notes

1. These 11 domains are: housing, income, jobs, community, education, environment, civic engagement, health, life satisfaction, safety and work–life balance; see http://www.oecdbetterlifeindex.org/, last accessed 17 January 2017.
2. While this is a more general point about the less than sharp boundaries between subjective and objective wellbeing, it is important to recognise that the conceptualisation of objective approaches to wellbeing more generally, and the selection and definition of universal human needs or capabilities more specifically, inevitably involves an element of normative judgement which cannot be wholly detached from concrete cultural and historical contexts. This has even been acknowledged by scholars working on "objective" needs, for instance, Max-Neef states that universal human needs do evolve very slowly over time, mentioning needs related to identity and freedom as examples of more recent developments (Guillen-Royo 2016: 45). Further complications arise if one is examining the ways in which objective needs can be measured. Again, scholars in this tradition have admitted that setting criteria, for instance to establish what "adequate" nutrition or housing should be, or which mental and physical states should be regarded as "healthy", are dependent on cultural norms, the state of science, etc. (Doyal and Gough 1991).
3. This not meant as a statement of resign (or even a pledge for flying, both authors support the #flyingless initiative which urges universities and professional associations to greatly reduce flying). We agree it is important to develop and maintain transnational academic networks. However, this should not require every participant to always be present in person as meetings, and even seminars and talks can nowadays be facilitated through remote video links (which also saves considerable costs).
4. The Gini coefficient is usually based on household income, not individual wages, and after tax, not before tax unlike the wage measure.

References

Alexander, S. 2012. Planned Economic Contraction: The Emerging Case for Degrowth. *Environmental Politics* 21 (3): 349–368.

Argyle, M. 1997. Is Happiness a Cause of Health? *Psychology & Health* 12 (6): 769–781.

Austin, A. 2016. Practical Reason in Hard Times: The Effects of Economic Crisis on the Kinds of Lives People in the UK Have Reason to Value. *Journal of Human Development and Capabilities* 17 (2): 225–244.

Bacigalupe, A., F.V. Shahidi, C. Muntaner, U. Martin, and C. Borrell. 2016. Why is there so Much Controversy Regarding the Population Health Impact of the Great Recession? Reflections on Three Case Studies. *International Journal of Health Services* 46 (1): 5–35.

Baxter, B. 1999. *Ecologism: An Introduction*. Washington, DC: Georgetown University Press.

Beckfield, J., C. Bambra, T.A. Eikemo, T. Huijts, C. McNamara, and C. Wendt. 2015. An Institutional Theory of Welfare State Effects on the Distribution of Population Health. *Social Theory & Health* 13 (3–4): 227–244.

Bentham, J., and R. Harrison. 1988. *Bentham: A Fragment on Government*. Cambridge: Cambridge University Press.

Blanchflower, D.G., and A.J. Oswald. 2004. Well-Being over Time in Britain and the USA. *Journal of Public Economics* 88 (7–8): 1359–1386.

Blewitt, J., and R. Cunningham. 2014. *The Post-Growth Project. How the End of Economic Growth Could Bring a Fairer and Happier Society*. London: Green House.

Bohk, C., and R. Rau. 2015. Impact of Economic Conditions and Crises on Mortality and its Predictability. *Kölner Zeitschrift Fur Soziologie Und Sozialpsychologie* 67: 271–294.

Borowy, I. 2013. Degrowth and Public Health in Cuba: Lessons from the Past? *Journal of Cleaner Production* 38: 17–26.

Brickman, P., and D.T. Campbell. 1971. Hedonic Relativism and Planning the Good Society. In *Adaptation-Level Theory*, ed. M.H. Appley, 215–231. New York: Academic Press.

Brickman, P., D. Coates, and R. Janoffbulman. 1978. Lottery Winners and Accident Victims—Is Happiness Relative. *Journal of Personality and Social Psychology* 36 (8): 917–927.

Cassiers, I. 2015. *Redefining Prosperity*. London: Routledge.

Catalano, R., and B. Bellows. 2005. Commentary: If Economic Expansion Threatens Public Health, Should Epidemiologists Recommend Recession? *International Journal of Epidemiology* 34 (6): 1212–1213.

Chang, S.S., D. Stuckler, P. Yip, and D. Gunnell. 2013. Impact of 2008 Global Economic Crisis on Suicide: Time Trend Study in 54 Countries. *BMJ-British Medical Journal* 347: f5239.

Daly, H., and J.B. Cobb. 1989. *For the Common Good. Redirecting the Economy Toward Community, the Environment, and a Sustainable Future*. Boston: Beacon Press.

Daly, H., and J. Farley. 2011. *Ecological Economics. Principles and Applications*, 3rd ed. Washington: Island Press.

Davies, W. 2015. *The Happiness Industry. How the Government and Big Business Sold us Wellbeing*. London/New York: Verso.

De Vogli, R., and J.T. Owusu. 2015. The Causes and Health Effects of the Great Recession: From Neoliberalism to "Healthy De-Growth". *Critical Public Health* 25 (1): 15–31.

Deaton, A. 2008. Income, Health, and Well-Being Around the world: Evidence from the Gallup World Poll. *Journal of Economic Perspectives* 22 (2): 53–72.

Deeming, C. 2013. Addressing the Social Determinants of Subjective Wellbeing: The Latest Challenge for Social Policy. *Journal of Social Policy* 42: 541–565.

Diener, E., and M.E.P. Seligman. 2004. Beyond Money: Toward an Economy of Well-Being. *Psychological Science in the Public Interest* 5 (1): 1–31.

Diener, E., and L. Tay. 2015. Subjective Well-Being and Human Welfare Around the World as Reflected in the Gallup World Poll. *International Journal of Psychology* 50 (2): 135–149.

Dolan, P., and R. Metcalfe. 2012. Measuring Subjective Wellbeing: Recommendations on Measures for Use by National Governments. *Journal of Social Policy* 41 (2): 409–427.

Dong, X., B. Milholland, and J. Vijg. 2016. Evidence for a Limit to Human Lifespan. *Nature* 538 (7624): 257–259.

Doyal, L., and I. Gough. 1991. *A Theory of Human Need*. Basingstoke: Palgrave.

Easterlin, R.A. 1974. Does Economic Growth Improve the Human Lot? In *Nations and Households in Economic Growth: Essays in Honor of Moses Abramovitz*, ed. P.A. David and M.W. Rede, 89–125. New York: Academic Press.

Easterlin, R.A., L.A. McVey, M. Switek, O. Sawangfa, and J.S. Zweig. 2010. The Happiness-Income Paradox Revisited. *Proceedings of the National Academy of Sciences of the United States of America* 107 (52): 22463–22468.

Easterly, W. 1999. Life During Growth. *Journal of Economic Growth* 4 (3): 239–276.

Elgin, D. 1982. *Voluntary Simplicity: An Ecological Lifestyle that Promotes Personal and Social renewal*. Toronto: Bantam Books.

Fanning, A.L. 2016. Policy Options for Sustainable and Equitable Coastal Economies: A Comparative Case Study. Doctoral Thesis, University of Cadiz.

Fleurbaey, M., and D. Blanchet. 2013. *Beyond GDP: Measuring Welfare and Assessing Sustainability*. New York: Oxford University Press.

Fritz, M., and M. Koch. 2014. Potentials for Prosperity Without Growth: Ecological Sustainability, Social Inclusion and the Quality of Life in 38 Countries. *Ecological Economics* 108: 191–199.

Fritz, M., and M. Koch. 2016. Economic Development and Prosperity Patterns Around the World: Structural Challenges for a Global Steady-State Economy. *Global Environmental Change* 38: 41–48.

Gavrilova, N.S., V.G. Semyonova, G.N. Evdokushkina, and L.A. Gavrilov. 2000. The Response of Violent Mortality to Economic Crisis in Russia. *Population Research and Policy Review* 19 (5): 397–419.

Gerdtham, U.-G., and C.J. Ruhm. 2006. Deaths Rise in Good Economic Times: Evidence from the OECD. *Economics & Human Biology* 4 (3): 298–316.

Gorz, A. 1980. *Ecology as Politics*. London: Pluto Press.

Gough, I. 2015. Climate Change and Sustainable Welfare: The Centrality of Human Needs. *Cambridge Journal of Economics* 39 (5): 1191–1214.

Granados, J.A.T. 2012. Economic Growth and Health Progress in England and Wales: 160 Years of a Changing Relation. *Social Science and Medicine* 74 (5): 688–695.

Granados, J.A.T., and E.L. Ionides. 2008. The Reversal of the Relation Between Economic Growth and Health Progress: Sweden in the 19th and 20th Centuries. *Journal of Health Economics* 27 (3): 544–563.

Guillen-Royo, M. 2016. *Sustainability and Wellbeing: Human Scale Development in Practice*. London; New York: Routledge, Taylor & Francis Group.

Habibov, N., and E. Afandi. 2015. Pre- and Post-crisis Life-Satisfaction and Social Trust in Transitional Countries: An Initial Assessment. *Social Indicators Research* 121 (2): 503–524.

Haw, C., K. Hawton, D. Gunnell, and S. Platt. 2015. Economic Recession and Suicidal Behaviour: Possible Mechanisms and Ameliorating Factors. *International Journal of Social Psychiatry* 61 (1): 73–81.

Heatley, B. 2014. Joined-Up Economics: The Political Economy of Sustainability, Financial Crises, Wages, Equality and Welfare. In *The Post-Growth Project. How the End of Economic Growth Could Bring a Fairer and Happier Society*, ed. J. Blewitt and R. Cunningham, 8–50. London: Green House.

Helliwell, J.F., H. Huang, and S. Wang. 2015. The Geography of World Happiness. In *World Happiness Report 2015(12–41)*, ed. J.F. Helliwell, R. Layard, and J. Sachs. New York: Sustainable Development Solutions Network.

Hertzman, C., and A. Siddiqi. 2000. Health and Rapid Economic Change in the Late Twentieth Century. *Social Science and Medicine* 51 (6): 809–819.

Inglehart, R. 1981. Post-Materialism in an Environment of Insecurity. *American Political Science Review* 75 (4): 880–900.

Jackson, T., and N. McBride. 2005. Measuring Progress? A Review of "Adjusted" Measures of Economic Welfare in Europe. CES Working Paper 11/05. Guildford: University of Surrey.

Jackson, T. 2011. *Prosperity without Growth: Economics for a Finite Planet*. London: Earthscan/Routledge.

Jackson, T., and P.A. Victor. 2016. Does Slow Growth Lead to Rising Inequality? Some Theoretical Reflections and Numerical Simulations. *Ecological Economics* 121: 206–219.

Jamison, D.T., L.H. Summers, G. Alleyne, K.J. Arrow, S. Berkley, A. Binagwaho, F. Bustreo, D. Evans, R.G.A. Feachem, J. Frenk, G. Ghosh, S.J. Goldie, Y. Guo, S. Gupta, R. Horton, M.E. Kruk, A. Mahmoud, L.K. Mohohlo, M. Ncube, A. Pablos-Mendez, K.S. Reddy, H. Saxenian, A. Soucat, K.H. Ulltveit-Moe, and G. Yamey. 2013. Global Health 2035: A World Converging within a Generation. *Lancet* 382 (9908): 1898–1955.

Kangas, O. 2010. One Hundred Years of Money, Welfare and Death: Mortality, Economic Growth and the Development of the Welfare State in 17 OECD Countries 1900–2000. *International Journal of Social Welfare* 19: S42–S59.

Karanikolos, M., P. Mladovsky, J. Cylus, S. Thomson, S. Basu, D. Stuckler, J.P. Mackenbach, and M. McKee. 2013. Financial Crisis, Austerity, and Health in Europe. *The Lancet* 381 (9874): 1323–1331.

Kasser, T. 2002. *The High Price of Materialism*. Cambridge, MA: MIT.

Kasser, T., and R.M. Ryan. 1993. A Dark Side of the American-Dream—Correlates of Financial Success as a Central Life Aspiration. *Journal of Personality and Social Psychology* 65 (2): 410–422.

Koch, M., and H. Buch-Hansen. 2016. Human Needs, Steady-State Economics and Sustainable Welfare, In *Sustainability and the Political Economy of Welfare*, ed. M. Koch and O. Mont, 29–43. London: Routledge.

Koch, M., H. Buch-Hansen, and M. Fritz. 2017. Shifting Priorities in Degrowth Research: An Argument for the Centrality of Human Needs. *Ecological Economics* 138: 74–81.

Kubiszewski, I., R. Costanza, C. Franco, P. Lawn, J. Talberth, T. Jackson, and C. Aylmer. 2013. Beyond GDP: Measuring and Achieving Global Genuine Progress. *Ecological Economics* 93: 57–68.

Lamb, W.F., J.K. Steinberger, A. Bows-Larkin, G.P. Peters, J.T. Roberts, and F.R. Wood. 2014. Transitions in Pathways of Human Development and Carbon Emissions. *Environmental Research Letters* 9 (1): 10.

Lane, R.E. 2000. *The Loss of Happiness in Market Democracies*. New Haven, CT: Yale University Press.

Lawn, P., and M. Clarke. 2010. The End of Economic Growth? A Contracting Threshold Hypothesis. *Ecological Economics* 69 (11): 2213–2223.

Layard, R. 2005. *Happiness: Lessons from a New Science*. London: Allen Lane.

Layard, R., G. Mayraz, and S. Nickell. 2010. Does Relative Income Matter? Are the Critics Right? In *International Differences in Well-being*, eds. E. Diener, J. F. Helliwell and D. Kahneman, 139–165. Oxford: Oxford University Press.

Marmot, M.G., G.D. Smith, S. Stansfeld, C. Patel, F. North, J. Head, I. White, E. Brunner, and A. Feeney. 1991. Health Inequalities Among British Civil-Servants—The Whitehall-II Study. *Lancet* 337 (8754): 1387–1393.

Max-Neef, M., A. Elizalde, and M. Hopenhayn. 1991. *Human Scale Development. Conception, Application and Further Reflections*. New York: The Apex Press.

Max-Neef, M. 1995. Economic Growth and Quality of Life: A Threshold Hypothesis. *Ecological Economics* 15 (2): 115–118.

McGuire, J.W. 2001. Social Policy and Mortality Decline in East Asia and Latin America. *World Development* 29 (10): 1673–1697.

McKee-Ryan, F., Z. Song, C.R. Wanberg, and A.J. Kinicki. 2005. Psychological and Physical Well-Being During Unemployment: A Meta-Analytic Study. *Journal of Applied Psychology* 90 (1): 53–76.

Mertens, A., and M. Beblo. 2016. Self-Reported Satisfaction and the Economic Crisis of 2007–2010: Or How People in the UK and Germany Perceive a Severe Cyclical Downturn. *Social Indicators Research* 125 (2): 537–565.

Muraca, B. 2012. Towards a Fair Degrowth-Society: Justice and the Right to a "Good Life" Beyond Growth. *Futures* 44 (6): 535–545.

Muraca, B., and T. von Egan-Krieger. 2011. Gerechtigkeit und gutes Leben jenseits von Wachstum. In *Ausgewachsen! Ökologische Gerechtigkeit, Soziale Rechte, Gutes Leben*, ed. W. Rätz, T. von Egan-Krieger, B. Muraca, A. Passadakis, M. Schmelzer, and A. Vetter, 43–56. Hamburg: VSA.

NEF. 2016. *Happy Planet Index 2016—Methods Paper*. London: New Economics Foundation.

Nussbaum, M.C. 2000. *Women and Human Development: The Capabilities Approach*. Cambridge: Cambridge University Press.

Nussbaum, M.C. 2003. Capabilities as Fundamental Entitlements: Sen and Social Justice. *Feminist Economics* 9 (2–3): 33 59.

Nussbaum, M.C. 2011. *Creating Capabilities: The Human Development Approach*. Cambridge, MA: Belknap.

O'Neill, J. 2008. Happiness and the Good Life. *Environmental Values* 17 (2): 125–144.

O'Neill, J. 2011. The Overshadowing of Needs. In *Sustainable Development*, ed. F. Rauschmayer, I. Omann, and J. Frühmann, 25–43. London: Routledge.

O'Neill, D. 2012. Measuring Progress in the Degrowth Transition to a Steady-State Economy. *Ecological Economics* 84: 221–231.

O'Neill, D. 2015. The Proximity of Nations to a Socially Sustainable Steady-State Economy. *Journal of Cleaner Production* 108: 1213–1231.

Odum, H.T., and E.C. Odum. 2006. The Prosperous Way Down. *Energy* 31 (1): 21–32.

Osberg, L., and A. Sharpe. 2002. An Index of Economic Well-Being for Selected OECD Countries. *Review of Income and Wealth* 3: 291–316.

Paech, N. 2013. *Befreiung vom Überfluss. Auf dem Weg in die Postwachstumsökonomie*. München: Oekom.

Page, E.A. 2007. Intergenerational Justice of What: Welfare, Resources or Capabilities? *Environmental Politics* 16 (3): 453–469.

Parsons, M.A. 2014. *Dying Unneeded: The Cultural Context of the Russian Mortality Crisis*. Nashville: Vanderbilt University Press.

Pieterse, J.N. 1998. My Paradigm or Yours? Alternative Development, Post-Development, Reflexive Development. *Development and Change* 29 (2): 343–373.

Piketty, T., and E. Saez. 2014. Inequality in the Long Run. *Science* 344 (6186): 838–843.

Posner, S.M., and R. Costanza. 2011. A Summary of ISEW and GPI Studies at Multiple Scales and New Estimates for Baltimore City, Baltimore County, and the State of Maryland. *Ecological Economics* 70 (11): 1972–1980.

Preston, S.H. 1975. Changing Relation between Mortality and Level of Economic-Development. *Population Studies-a Journal of Demography* 29 (2): 231–248.

Pritchett, L., and M. Viarengo. 2010. Explaining the Cross-National Time Series Variation in Life Expectancy: Income, Women's Education, Shifts, and What Else? Human Development Research Paper United Nations Development Programme (31).

Rahnema, M., and V. Bawtree. 1997. *The Post-Development Reader*. London: Zed Books.

Research & Degrowth. 2010. Degrowth Declaration of the Paris 2008 Conference. *Journal of Cleaner Production* 18 (6): 523–524.

Robeyns, I. 2005. The Capability Approach: A Theoretical Survey. *Journal of Human Development* 6 (1): 93–117.

Rosa, H., and J. Trejo-Mathys. 2013. *Social Acceleration: A New Theory of Modernity*. New York: Columbia University Press.

Rosero-Bixby, L., and W.H. Dow. 2016. Exploring Why Costa Rica Outperforms the United States in Life Expectancy: A Tale of Two Inequality Gradients. *Proceedings of the National Academy of Sciences of the United States of America* 113 (5): 1130–1137.

Ruhm, C.J. 2000. Are Recessions Good for Your Health? *Quarterly Journal of Economics* 115 (2): 617–650.

Ryan, R.M., V. Huta, and E.L. Deci. 2008. Living Well: A Self-Determination Theory Perspective on Eudaimonia. *Journal of Happiness Studies* 9 (1): 139–170.

Schneider, F., G. Kallis, and J. Martinez-Alier. 2010. Crisis or Opportunity? Economic Degrowth for Social Equity and Ecological Sustainability. Introduction to this Special Issue. *Journal of Cleaner Production* 18 (6): 511–518.

Sen, A. 1999. *Development as Freedom*. Oxford: Oxford University Press.

Soper, K. 2008. Alternative Hedonism, Cultural Theory and the Role of Aesthetic Revisioning. *Cultural Studies* 22 (5): 567–587.

Stearns, P.N. 2012. The History of Happiness How the Pursuit of Contentment has Shaped the West's Culture and Economy. *Harvard Business Review* 90 (1–2): 104–109.

Steinberger, J.K., and J.T. Roberts. 2010. From Constraint to Sufficiency: The Decoupling of Energy and Carbon from Human Needs, 1975–2005. *Ecological Economics* 70 (2): 425–433.

Stevens, A.H., D.L. Miller, M.E. Page, and M. Filipski. 2015. The Best of Times, the Worst of Times: Understanding Pro-Cyclical Mortality. *American Economic Journal-Economic Policy* 7 (4): 279–311.

Suhrcke, M., and D. Stuckler. 2012. Will the Recession be Bad for Our Health? It Depends. *Social Science and Medicine* 74 (5): 647–653.

Sullivan, D., and T. von Wachter. 2009. Job Displacement and Mortality: An Analysis Using Administrative Data. *The Quarterly Journal of Economics* 124 (3): 1265–1306.

Szreter, S. 1999. Rapid Economic Growth and 'The Four Ds' of Disruption, Deprivation, Disease and Death: Public Health Lessons from Nineteenth-Century Britain for Twenty-First-Century China? *Tropical Medicine & International Health* 4 (2): 146–152.

Talberth, J., C. Cobb, and N. Slattery. 2007. *The Genuine Progress Indicator 2006: A Tool for Sustainable Development*. San Francisco: Redefining Progress.

Tversky, A., and D. Kahneman. 1991. Loss Aversion in Riskless Choice: A Reference-Dependent Model. *The Quarterly Journal of Economics* 106 (4): 1039–1061.

Victor, P.A. 2012. Growth, Degrowth and Climate Change: A Scenario Analysis. *Ecological Economics* 84: 206–212.

Victor, P.A., and G. Rosenbluth. 2007. Managing Without Growth. *Ecological Economics* 61 (2–3): 492–504.

Wilkinson, R.G., and K.E. Pickett. 2009. *The Spirit Level. Why More Equal Societies Almost Always Do Better*. London: Allen Lane.

Wolf, C. 2009. Intergenerational Justice, Human Needs, and Climate Policy. *In A Intergenerational justice*, ed. A. Gosseries and L.H. Meyer, 347–376. Oxford: Oxford University Press.

Zivin, K., M. Paczkowski, and S. Galea. 2011. Economic Downturns and Population Mental Health: Research Findings, Gaps, Challenges and Priorities. *Psychological Medicine* 41 (7): 1343–1348.

CHAPTER 6

The Generation of Human Wellbeing: Social Practices Theory

Abstract This chapter proposes that social practices theory can provide us with important insights for conceptualising possible wellbeing implications of postgrowth. Highlighting the interaction between "structure" and "agency" in the generation of wellbeing, this chapter reflects on the ways in which different social dimensions—e.g. resources, institutions and discourses—interact in generating wellbeing at individual and social levels, and the roles that stability and change play in this process. An important question emerging from this is whether the various social structures that are currently organised around market capitalism and its inherent power structures can change rapidly enough, and at similar speeds, to avoid deteriorations of wellbeing, including the important role of cultural change.

Keywords Social practices · Structure and agency · Micro and macro Power · Wellbeing

To advance our thinking about potential implications of postgrowth for wellbeing, we need a theory of the ways in which wellbeing is generated and how it is impacted by social change generally and economic contraction specifically. This chapter explores ways in which we can utilise social practice theory to conceptualise the generation of wellbeing and the relationship between social change/economic contraction and wellbeing. Social practice theory provides us with a more sociologically grounded

M. Büchs and M. Koch, *Postgrowth and Wellbeing*,
DOI 10.1007/978-3-319-59903-8_6

understanding of the generation of wellbeing because it pulls both macro- and micro-dimensions of the social into view and focuses on how they constitute each other. This is important as it suggests that wellbeing is co-generated by both dimensions: it does simultaneously emerge from individuals' perceptions and actions as well as from the contexts in which social activity is embedded. This has several implications for understanding wellbeing which we will discuss in this chapter. The social practices lens is also useful for the purpose of thinking about possible impacts of (rapid) socio-economic change on wellbeing because at its centre stand questions about how stability and change emerge in society. The chapter will start by providing a brief overview of social practice theory, before connecting it to the discussion of wellbeing.

Social Practices Theory: A Brief Overview

Social practice theory has a long history in sociology and draws on a variety of different approaches. It responds to a central problem in social theory which concerns the relationships between "micro" (including "subjective" perceptions, minds, identities, values, etc.) and "macro" ("objective" structures or systems) dimensions of society. Numerous sociologists have contributed to conceptualising the macro–micro or subject–object relationship, including Marx' thinking on the connections between economic categories, corresponding social structures and relations, as well as modes of consciousness and agency (Marx 1961, 2006); Parsons' *Structure of Social Action* (1968), Bourdieu's "praxeology" (1977, 1990), Giddens' theory of structuration (1984) and Habermas' theory of communicative action (1984, 1987).

More recently, several scholars have sought to integrate various strands of thinking about micro–macro relationships and declared a "practice turn" in sociology (Schatzki et al. 2001). There are several versions of more recent social practices accounts, but all of them have in common that they locate the intersection between micro- and macro-dimensions of society in social practices which are understood as a constant flow of performances of "doings and sayings" (Schatzki 1996). Social practices are thus declared as the "smallest [and hence primary] unit of social analysis" (Reckwitz 2002: 249).

For the purpose of this book, we draw on and combine ideas on social practices developed by a variety of scholars, including Giddens (1984), Bourdieu (1977, 1990), Shove (2012), Schatzki (1996) and Mouzelis

(2008). Following Mouzelis (2008: 226–227), we do not regard micro- and macro-dimensions of society as ontologically separate. Instead, the distinction can be understood as a form of "methodological dualism" which enables us to examine and understand social processes from both of these perspectives and to conceptualise the ways in which they are coupled to each other through social practices. While social practices are the performances—sayings and doings—carried out by individual and collective actors, they simultaneously draw on and reproduce "structured" contexts. Often, social structures are understood to be located at the macro-level of society. We argue here that this conception of structures is misleading because "structured dimensions" of social practices are relevant at both macro- and micro-levels of society. According to this understanding, examples for structured dimensions at the macro-level include social discourses, culture and institutions (formal laws, and informal rules and norms), and examples for structured dimensions at the micro-level comprise people's worldviews, identities, and competences. All of these phenomena, both at the micro- and the macro-level, can be regarded as structural because they have "ordering", stabilising characteristics. While, as argued above, they can be distinguished analytically, they are closely interlinked in reality as individuals' worldviews and values simultaneously make up and are shaped by collective discourses and cultures. These macro- and micro-structures form a range of social fields, including the economic system—currently growth-based market capitalism—political and legal systems—, cultures, the welfare state, education and academia, and so forth. These social fields are closely coupled to each other in the sense that in any society, the organisation of any one field shapes that of all of the other fields and vice versa. While structures exist at both macro- and micro-levels, it needs to be emphasised that actors inherently possess agency—the capacity to act and to do so creatively and in novel ways. Human agency is where social change and hence a change of structural properties of social practices originates.

While older accounts of social practices tended to concentrate on conceptualising the relationship between macro- and micro-dimensions of society, more recent accounts of social practice theory have made an important contribution by including (manufactured) "things" and "infrastructures" into the conception of social practices, drawing on actor–network theory (Latour 2005). For instance, for Shove (2012), things and infrastructures are one of the elements that constitute social practices, in addition to meanings and skills. Less clear in this debate is the status of "ecological contexts" in the conception of social practices, which can be understood to include ecosystems, natural resources, environmental space at the macro- as well as human bodies and their functioning at the micro-level. The more recent accounts of social practices which regard things, technologies and infrastructures as elements of social practices (Shove et al. 2012), do not usually include "ecological contexts" and instead regarded them as external to practices. We argue here that while ecosystems are indeed partly independent from social practices, they remain to be co-constituted or shaped by practices (e.g. resources are extracted, used up, disposed of; natural habitats and our bodies are impacted by social practices). Environmental dimensions of society also provide important, and unequally distributed, resources and contexts for practices—an example is the ways in which the extraction, production and use of specific types of resources such as iron, bronze, steal, coal and oil have shaped the organisation of societies in different phases of history, and the ways in which our disposal of greenhouse gas emissions in the atmosphere (and our responses to the consequences) is going to impact on societies for a very long time to come. It is hence impossible to understand social practices without considering the role of these environmental dimensions. Therefore, we argue here that these "ecological contexts" should be included in social practice frameworks. For this, we can draw on co-evolutionary approaches which conceptualise possible pathways for low-carbon transitions and regard "ecosystems", along with "technologies", "institutions" and "practices" (by users and businesses) as co-evolving systems (Foxon 2011). What our framework adds is that we integrate the core idea from social practice theory about the co-constitutive relationship between practices (by various types of actors and organisations at different levels) on the one hand, and social, technological and ecological systems on the other.

When we speak of "structured dimensions" of practices, "structured" has two meanings, first, as briefly mentioned above, it refers to something that is relatively stable over time, in the sense that these dimensions are repeatedly being drawn on and thus reproduced through the performance of practices. Furthermore, social practices are stabilised because the various dimensions involved in them—here conceptualised as socio-eco-technological structures—become linked to each other (Shove et al. 2012)—practices are made up of specific, closely tied, constellations of discourses, institutions, technologies, identities, competences, ecosystems, etc.

In addition, we include here a second meaning of "structured" to highlight that these various dimensions of practices manifest the unequal distribution of resources over space, time and social networks and that these patterns are reproduced through the performance of social practices. Resources is a broad term here to include material, financial and cultural resources as well as power. What thus emerges are patterns of dominant and more marginal practices, where dominant practices draw on greater and more powerful material and immaterial resources and form more dominant socio-eco-technological regimes.

All of this has implications for the postgrowth debate, and as we will see in the next section, for thinking about the wellbeing implications of transitioning beyond a growth-based society. One important implication is that market capitalism—and economic growth that is driven by it—has structural properties and is made up of and shapes a range of tightly coupled structures, including institutions, norms, discourses, culture, technologies, competences, identities and ecosystems. Every action that contributes to the performance of a social practice, for instance the purchase of a cup of coffee as part of wider food practices, draws on and reproduces structural features of growth-dependent market capitalism, including the environmental implications and global social interdependencies and inequalities of coffee production and consumption. As numerous sociologists have discussed, the reproduction of the social system through practices often remains independent from people's intentions—the primary intention of the person purchasing the cup of coffee may be to enjoy the taste of coffee and its energising effect, while the wider structural dimensions that this action reproduces often stay out of view (e.g. see Giddens' (1984: 8–12) discussion of unintended consequences of intentional action or Bourdieu's (1977: 164) account of the reproduction of social structures through "taken for granted", "undisputed" understandings). From this perspective, economic growth is not just an external premise that actors can decide to act upon or not, but it is a principle with structural properties that is engrained in ways of thinking and acting—for the most part habitually. Growth thus becomes something that is perceived as "natural" by the vast majority of actors in the economy—something that we illustrated using examples from Marx in Chap. 2 who demonstrated this process in relation to the profit motive and workers' motivation to participate in the labour market. With Welzer (2011), it can be argued that the growth paradigm currently is at the core of our "mental infrastructures" in that people's identities and life

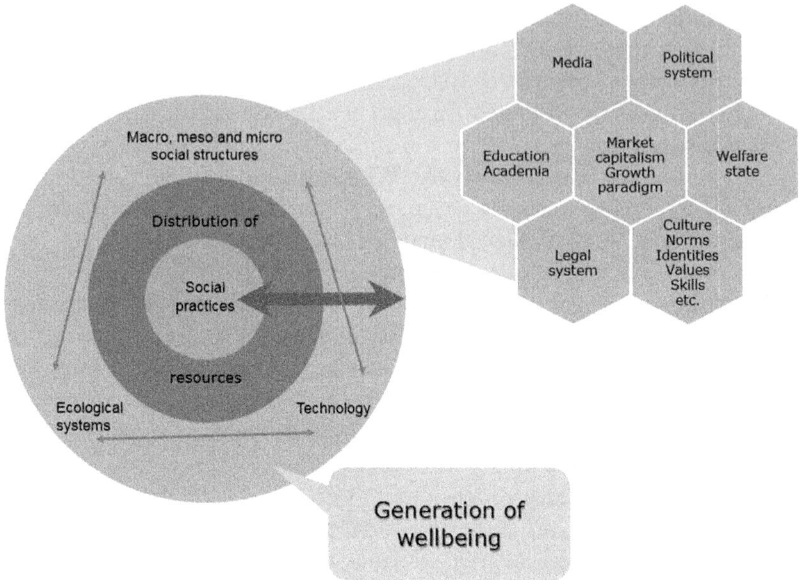

Fig. 6.1 Social practices, coupled social structures and the generation of wellbeing outcomes

goals are closely aligned with it—shaped by ideas of social progress, personal status and success through careers, rising income and consumption. Even seemingly alternative goals such as "personal fulfilment" can be infused with ideas that remain tightly coupled to the growth paradigm, for instance, if fulfilment is sought through high consumption and high emissions practices such as extensive long haul travel or expensive hobbies and gadgets (Fig. 6.1).

Wellbeing Through a Social Practices Lens

How then can the social practices framework outlined above help us to think about possible implications of postgrowth for wellbeing? First, this lens offers the insight that wellbeing is generated through social practices. From this follows, second, the implication that to transition away from the growth paradigm, a range of coupled social structures need to change simultaneously. Different speeds of change of different social structures may have implications for the generation of wellbeing.

The first insight that wellbeing is generated through social practices fits well with recent developments in the literature on health and wellbeing. For instance, Maller (2015: 54) regards "health and wellbeing [as] (...) outcomes of participation in a set of social practices". In this conception, the focus lies on the wellbeing of individuals (who "participate" in specific social practices). We take a broader view here to include wellbeing outcomes at the collective level. In this sense, wellbeing outcomes at individual and collective levels depend on the character of social practices and implicated social structures that are predominant in society. One example is the relationship between culture and wellbeing: the generation of wellbeing through social practices implies the continuous negotiation and (re-)establishment of conceptions of wellbeing goals—which are closely related to culturally dominant values and life goals. Wellbeing conceptions are thus part of both macro-level discourses and micro-level worldviews and meanings. It is also at these levels that societal communication and negotiations take place about the culturally specific ways in which wellbeing goals are going to be satisfied. A second example is that the achievement of wellbeing goals through the performance of social practices also depends on the character of the institutions that are implied in social practices, including the ways in which work, welfare, health and education are governed, as well as on the structure of the distribution of a range of resources that support health and wellbeing.

The social practices lens also contributes to thinking about possible wellbeing implications of social change generally and a transition away from a growth-based economy specifically. To be able to develop some of these thoughts, it is instructive to review the ways in which social change is conceptualised in social practice theory. Even though the micro- and macro-contexts of social practices take on structured qualities which stabilise practices, social practice theorists have been keen to emphasise that the possibility of change is inherent in the concept of practices. First of all, practices are performances which continuously need to be re-enacted. This in itself is thought to open up possibilities for diversion from trodden paths. Actors' ability to question, reflect and creatively search for new ways of saying and doing things means that they can always change the ways in which practices are performed by drawing on different or new meanings, norms, worldviews, identities, competences, resources, etc. Reflection—the becoming aware of, questioning and criticising of existing practices—is thus a precondition for change. Some sociologists have suggested that the questioning of the "taken for granted" which brings the "unformulated into formulation" (Bourdieu 1977: 168) and then opens up the opportunity of change often occurs in moments of crisis, in which subjective and objective structures no longer fit (ibid.: 168–169; see Giddens 1984: 61–64).

A change of social practices also involves a breaking up of the links between the structural dimensions of practices. While some of the "old" dimensions can continue to be involved in a new social practice, fresh links need to be established between them and the other dimensions of practices. An example is the way in which in the early phases of emerging practices of car driving, competences of steering and repairing a vehicle needed to be adapted from horse riding and combined with existing meanings of exhibiting wealth, enjoying adventure and nature (Shove et al. 2012: 26–29).

While the possibility of change is thus inherent in the concept of social practices, the stabilisation of some dimensions of practices over time, at least relatively speaking, is equally part of the social practice concept. As discussed above, the existence of practices implies the formation of close links between their constituent structures. For example, the performance of practices that involve certain technologies is also attached to certain competences, norms and worldviews which become tightly coupled and hence more difficult to shift away from, as has been demonstrated using examples of car-based mobilities (Urry 2004) or the invention of showering (Shove 2003). Stability of some dimensions of social practices provides orientation and an extent of predictability of how oneself and other people are going to act in the future, providing a framework within which flexibility is possible. This orienting function of some dimension of practices is likely to be an important condition for people to form identities and relationships—key ingredients for wellbeing.

The orienting role of social structures might explain why some research has found that across 174 countries, several measures of wellbeing and social performance, including life satisfaction, health, safety and trust, voice and accountability, were highest in periods of economic stability, but lower in times of GDP growth or contraction (O'Neill 2015); or that life expectancy can be negatively affected by both rapid economic growth and contraction (Notzon et al. 1998; Szreter 1999). It also connects with a thought expressed by Wolfgang Streeck who predicts that capitalism will come to an end as a result of three mutually reinforcing trends—declining growth, rising inequality, and rising public and private debt. His notes about possible implications of societies that find themselves in such a phase of fundamental transition ring many alarm bells for considerations about future wellbeing:

> "The breakdown of a social order in the absence of a successor may give rise to (...) a society devoid of coherent institutions capable of normalizing the lives of its members and protecting them from accidents and monstrosities of all sorts. Life in an interregnum is characterized by a lack of structural determinacy, making it unpredictable. Its society fails to provide its members with reliable templates around which they may organize themselves: instead it demands constant improvisation, making individuals substitute strategy for structure—offering rich opportunities to oligarchs and warlords of all sorts while forcing the majority to live in insecurity, uncertainty and anomy". (Streeck et al. 2016: 169)

A further important issue that might explain the stabilisation of some dimensions of social practices concerns the role of power. As mentioned above, one of the meanings that relate to the idea that social practices draw on and constitute social structures is that resources are unequally distributed in the performance of practices—and often very stably so. This includes the unequal distribution of wealth, income, knowledge, worldviews, competences, as well as access to technologies, infrastructures and social networks. The unequal distribution of these resources establishes unequal power positions in society where those who benefit from the set-up of the current system tend to be more powerful and will hence have an interest and likely greater capacity to prevent the system from changing. The stabilisation of these layers of social practices can sometimes make their change "from the bottom-up" challenging. One example relates to the role of the corporate media in the 2016 US election campaign, where the three American television network broadcasts ABC World News, CBS Evening News and NBC Nightly News gave Trump more than twice as much coverage (1144 stories) as Clinton (506 stories) and nearly ten times more than Bernie Sanders (115 stories) (Tyndall Report 2016). Financial, political and discursive power merged to create a context which lastingly shaped voter behaviour.

What then are implications of ideas that social practices draw on and establish closely coupled structural dimensions and that the orientation that some dimensions of practices provide might support wellbeing? The first idea of closely coupled structural dimensions of social practices implies that for a smooth transition from one formation of social practices to another to happen, the different structural dimensions of social practices would simultaneously need to change and reconnect to each other. The question then becomes under which circumstances this is likely or unlikely to happen. What is clear is that it cannot be taken for granted that structural dimensions change with similar speeds in phases of rapid transition. For instance, we can imagine a situation in which the economy and the goods and services it provides transform rapidly when growth halts or contracts, supply shortages emerge, important infrastructures are interrupted through extreme weather events or power failures, while social norms, discourses, worldviews and other institutions respond more slowly.

The question then becomes how unequal speeds of change of constituent social structures in phases of social transition might impact on individual and collective wellbeing. Some examples from sociological and psychological research suggest that different speeds of changing social structures can establish misalignments and disruptions of social practices which can in turn negatively influence health and other wellbeing outcomes. For instance, in a classical study, Durkheim presents suicide at least partly as an outcome of a failure of cultural resources to provide meaning and orientation in the context of other, more rapid social changes (Durkheim 2006; Vega and Rumbaut 1991: 375). Another relevant idea for this discussion has been formulated by Bourdieu with what he called the "hysteresis effect". Here, Bourdieu emphasises that, especially during phases of social transition, people's habitus and "objective" social circumstances can become disjointed: as a result of hysteresis, dispositions can be "out of line with the field and with the "collective expectations" which are constitutive of its normality. This is the case, in particular, when a field undergoes a major crisis and its regularities (even its rules) are profoundly changed" (Bourdieu 2000: 160). This can contribute to a deterioration of people's wellbeing as it makes them feel "out of place" or let them be perceived that way and "plunges them deeper into failure" (Bourdieu 2000: 161) because they cannot make use of new opportunities or are mistreated or socially excluded by others. Empirical research which partly builds on the idea of hysteresis has shown that wide-ranging organisational change can have a range of negative effects on people's health and mortality (Ferrie et al. 1998; McDonough and Polzer 2012).

All this suggests that for wellbeing outcomes to be maintained or even improved in the context of postgrowth will rely on the emergence of new practices—and simultaneous transformation and new coupling of a variety of social structures implicated in practices. One example is the required change of dominant cultures, worldviews and identities which are currently rooted in the growth paradigm as discussed in the previous section. This is important because these cultural structures not only firmly establish the growth paradigm in economic theory and practice (Davey 2015), but also encompass socially dominant conceptions and practices of wellbeing. With Welzer (2011), it can be argued that the difficulties associated with the transformation of "mental infrastructures" have so far been underestimated in the postgrowth literature. This transformation will require a fundamental reorientation of culture and embedded wellbeing conceptions towards the fulfilment of basic human needs, framed by concern and care for the wellbeing of future generations. This seems like a utopian vision from the point of view of current growth-oriented consumer culture, and the question is whether such cultural change can develop fast enough to align with the required changes of scales of resource use, pollution output and associated institutional changes. Daly and Farley (2011: 12) remind us that social change has been extremely slow for most of human history—sufficiently slow for culture to adjust to these changes. Worryingly, they point out: "most likely we will have to change our cultural institutions and values in response, particularly the economic institutions and values that have led to this state of affairs. Since there is certainly some limit to how fast we can adapt culturally, we need also to consider seriously how to slow down the rate of change that is forcing the adaptations. It is worth remembering that not all change is desirable and that even desirable change can be too fast" (Daly and Farley 2011: 12; also see Polayi 1944). If these concerns are valid, our capacity to achieve the necessary societal transformations within required timescales without compromising current (and near future) generations' wellbeing is likely to be challenged.

While the emphasis here on the need for cultural change that is in alignment with a postgrowth society is just one example out of several other structural dimensions of practices that would need to change, it is likely to be an important one as it connects to a range of other structural dimensions. For instance, cultural change is also required for institutions (norms and policies) to be established that ensure a fairer distribution of wealth, resource use and emissions within and across countries, as well as across generations.

Conclusion

This chapter started with the recognition that existing evidence on the wellbeing implications of economic growth and contraction remains limited because they cannot factor in the institutional changes that a transition to sustainable postgrowth would involve. It argues that social theory can still provide us with some more general insights into possible wellbeing implications of far-reaching and rapid socio-economic change. Social practice theory is especially promising in this regard as it can show two things: first a recognition that wellbeing is generated through social practices which imply a variety of "structured dimensions", both at the level of the actor and of society, currently organised around market capitalism; and second that there is a tendency for these various structured dimensions to form links over time which provide orientation—a framework within which flexibility is possible. Wider social change that goes beyond incremental adjustments of our daily routines means that these links become unhitched, questioning and disrupting social structures. If alternative cultural repertoires, institutions, etc., cannot be established quickly, this might result in a loss of identity, orientation and social cooperation with likely impacts on wellbeing. In the next chapter, we discuss regulatory principles and levels, institutions and policy initiatives that might support a smoother transition towards sustainable postgrowth.

References

Bourdieu, P. 1990. *The Logic of Practice*. Cambridge: Polity Press.

Bourdieu, P. 1977. *Outline of a Theory of Practice*. Cambridge: Cambridge University Press.

Bourdieu, P. 2000. *Pascalian Meditations*. Cambridge: Polity Press in association with Blackwell Publishers.

Daly, H., and J. Farley. 2011. *Ecological Economics. Principles and Applications*, 3rd ed. Washington: Island Press.

Davey, B. 2015. *Credo: Economic Beliefs in a World in Crisis*. Dublin: FEASTA.

Durkheim, É. 2006. *On Suicide*. London: Penguin Books.

Ferrie, J.E., M.J. Shipley, M.G. Marmot, S. Stansfeld, and G.D. Smith. 1998. The Health Effects of Major Organisational Change and Job Insecurity. *Social Science and Medicine* 46 (2): 243–254.

Foxon, T.J. 2011. A Coevolutionary Framework for Analysing a Transition to a Sustainable Low Carbon Economy. *Ecological Economics* 70 (12): 2258–2267.

Giddens, A. 1984. *The Constitution of Society: Outline of the Theory of Structuration*. Cambridge: Polity Press.

Habermas, J. 1984. *The Theory of Communicative Action, vol. 1, Reason and the Rationalization of Society*. Boston: Beacon.

Habermas, J. 1987. *The Theory of Communicative Action, vol. 2, Lifeworld and System: A Critique of Functionalist Reason*. Boston: Beacon.

Latour, B. 2005. *Reassembling the Social. An Introduction to Actor-network Theory*. Oxford: Oxford University Press.

Maller, C.J. 2015. Understanding Health Through Social Practices: Performance and Materiality in Everyday Life. *Sociology of Health & Illness* 37 (1): 52–66.

Marx, K. 1961. *Capital: A Critique of Political Economy*, vol. 1. Moscow: Foreign Languages Publishing House.

Marx, K. 2006. *Capital: A Critique of Political Economy*, vol. 3. London: Penguin Classics.

McDonough, P., and J. Polzer. 2012. Habitus, Hysteresis, and Organizational Change in the Public Sector. *Canadian Journal of Sociology-Cahiers Canadiens De Sociologie* 37 (4): 357–379.

Mouzelis, N.P. 2008. *Modern and Postmodern Social Theorizing. Bridging the Divide*. Cambridge: Cambridge University Press.

Notzon, F.C., Y.M. Komarov, S.P. Ermakov, C.T. Sempos, J.S. Marks, and E.V. Sempos. 1998. Causes of Declining Life Expectancy in Russia. *Journal of the American Medical Association* 279 (10): 793–800.

O'Neill, D. 2015. The Proximity of Nations to a Socially Sustainable Steady-State Economy. *Journal of Cleaner Production* 108: 1213–1231.

Reckwitz, A. 2002. Towards a Theory of Social Practices: A Development in Culturalist Theorizing. *European Journal of Social Theory* 5 (2): 243–263.

Schatzki, T.R. 1996. *Social Practices. A Wittgensteinian Approach to Human Activity and the Social*. Cambridge: Cambridge University Press.

Schatzki, T.R., K. Knorr Cetina, and E. von Savigny. 2001. *The Practice Turn in Contemporary Theory*. London: Routledge.

Shove, E. 2003. *Comfort, Cleanliness and Convenience. The Social Organization of Normality*. Berg: Oxford.

Shove, E., M. Pantzar, and M. Watson. 2012. *The Dynamics of Social Practice. Everyday Life and How It Changes*. London: Sage.

Streeck, W., C. Calhoun, P. Toynbee, A. Etzioni, W. Streeck, C. Calhoun, P. Toynbee, and A. Etzioni. 2016. Does Capitalism have a Future? *Socio-Economic Review* 14 (1): 163–183.

Szreter, S. 1999. Rapid Economic Growth and 'The Four Ds' of Disruption, Deprivation, Disease and Death: Public Health Lessons from

Nineteenth-Century Britain for Twenty-First-Century China? *Tropical Medicine & International Health* 4 (2): 146–152.

Tyndall Report. 2016. *Year in Review 2016*. http://tyndallreport.com/yearinreview2016. Accessed 12 Feb 2017.

Urry, J. 2004. The "System" of Automobility. *Theory Culture & Society* 21 (4–5): 25–39.

Vega, W.A., and R.G. Rumbaut. 1991. Ethnic Minorities and Mental Health. *Annual Review of Sociology* 17 (1): 351–383.

Welzer, H. 2011. *Mentale Infrastrukturen. Wie das Wachstum in die Welt und in die Seelen kam*. Berlin: Heinrich-Böll Stiftung.

CHAPTER 7

Welfare in a Global Steady-State Economy

Abstract This chapter asks which kinds of institutions could support wellbeing in the context of sustainable postgrowth, especially in periods of far-reaching social change. It discusses core principles for the achievement of a steady-state economy—including a sustainable scale of material throughput and social equality, followed by considerations of the role of spatial scales—the requirement to share and coordinate responsibility for transitions towards postgrowth across global, national, regional and local levels. It also considers a range of policy proposals that have been made to support wellbeing under postgrowth, focusing on macroeconomic steering, inequality/redistribution, minimum and maximum incomes, carbon rationing, consumption, working time reduction, work–life balance as well as population/migration.

Keywords Steady-state economy · Sustainable scale · Multi-level governance · Redistribution · Carbon caps

Previous chapters have demonstrated that any re-embedding of economy and society into the environmental limits of the planet is unlikely as long as the top priority of economic growth in policy making is upheld. Yet it is far from obvious how economy and society in a postgrowth period could look like. Given the vast complexity involved in the transition to an economy that is not anymore primarily steered by monetary growth—a range of social systems would need to change in tandem

M. Büchs and M. Koch, *Postgrowth and Wellbeing*,
DOI 10.1007/978-3-319-59903-8_7

(Chap. 6)—it is difficult to foresee how these systems may interact. Discussing and outlining some of the general features of a postgrowth economy and society as well as of the associated welfare institutions is accordingly difficult and to some extent speculative. We would nevertheless argue that such forward thinking is required to encourage debate and eventually social change. In this last chapter, we sketch out what we see as key principles of a "global steady-state economy" (SSE) and an associated "sustainable welfare" system. Based on current debates in ecological economics and related social sciences, we start with a discussion of core principles of a global SSE, particularly a sustainable level of material throughput and social equality. This is followed by considerations of the role of spatial scales—the requirement to share and coordinate responsibility for transitions towards postgrowth across global, national, regional and local levels. The chapter concludes with a brief discussion of "eco-social" policies that may facilitate the transition to steady-state and sustainable welfare principles in rich countries.

Principles of a Global Steady-State Economy and Sustainable Welfare

Within ecological economics, Herman Daly's SSE (1972) is the most-cited concept of a non-growing economic system that functions within ecological boundaries (Chap. 5). We use its core principles as a point of departure for this last chapter. Basically, in an SSE the economic process is not primarily regarded from the standpoint of monetary growth but seen as a bio-physical process. It is a model of an economy that does not grow in the sense that it keeps the level of "throughput" within a sustainable scale, where throughput is defined as the "extraction of raw materials from nature and their return to nature as waste" (Farley 2013: 49). It is further assumed that in an SSE, continued technological advances in combination with shorter working hours will facilitate the maintenance of relatively high living standards with relative low resource consumption and carbon emissions (Jackson and Victor 2011; Mont 2016).

The original concept of an SSE was not developed at the global level. Yet environmental threats such as climate change are global issues, because for the atmosphere it does not matter from which part of the globe greenhouse gases are emitted. Accordingly, the ecological footprint and the associated matter and energy throughput of the whole planet would need to shrink if the world's mode of production and consumption were to respect ecological limits. However, due to massive differences in economic development and unprecedented socio-economic global inequality (Milanovic 2005), such a re-embedding of the world's production and consumption patterns would imply different challenges across regions and nations. Recent comparative research (O'Neill 2015; Fritz and Koch 2016) demonstrates that not only nations' social inclusion, wellbeing and democracy scores largely increase with GDP per capita but also their ecological footprints and carbon emissions. According to Fritz and Koch (2016), who divided 138 countries into five clusters of economic development measured as GDP per capita ("poor", "developing", "emerging", "rich" and "overdeveloped" countries), it is only the poorest group of countries that could currently be seen as environmentally sustainable (Fig. 7.1).

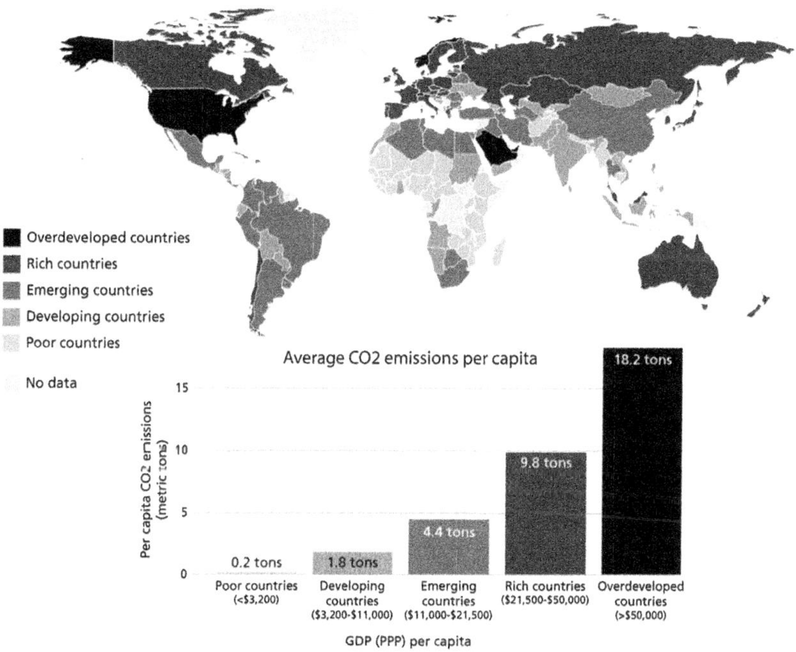

Fig. 7.1 The World's CO_2 emissions by level of economic development. *Source* Based on data in Fritz and Koch (2016). Detailed information on cluster composition and social, environmental and individual welfare and prosperity indicators for each country are provided in appendix to the article. We would like to express our thanks to Aron Strandberg for designing the figure and giving us permission to publish it

Such a global map of eco-social performances relative to GDP per capita has repercussions for any model of a global SSE as countries at different levels of economic development would need to undertake different measures to achieve maximum wellbeing within ecological limits (Fritz and Koch 2016: 48). The policy challenge for "poor" countries would be to enhance the quality of life and social inclusion while maintaining low ecological footprints and carbon emissions; "developing" and "emerging" countries face the double challenge of achieving individual wellbeing and social inclusion while preserving relatively low amounts of matter and energy throughput as well as carbon emissions[1]; "rich" and especially the "overdeveloped" countries would need to make production and consumption processes more environmentally sustainable to reduce implied amounts of material resources and fossil energy while maintaining welfare and wellbeing. The result that there would be different national and regional trajectories *en route* to a global SSE reflects earlier research by the degrowth-research community (O'Neill 2012: 221; Martínez-Alier et al. 2010).

We draw two preliminary conclusions from this. The first is that the socio-ecological transitions required for setting up a global SSE and associated sustainable welfare systems would involve issues of redistribution of wealth, labour, time and natural resources both within and between countries. Hence, any institutional welfare compromise related to an SSE would need to go beyond the national scale, at which post-war welfare arrangements were agreed (Chap. 3), and encompass the entire globe. We will come back to this in the following two sections. The second conclusion follows from the fact that the "developing" countries assembled in the second poorest cluster (Fritz and Koch 2016) already work and live, so to speak, beyond their ecological means while action to mitigate climate change would need to take place within an extremely short time period (IPCC 2014). In this situation, it is for the time being difficult to see how the provision of sustainable welfare for all present and future inhabitants of the world could mean much more than the satisfaction of basic human needs. Or, put differently, the amount of societal wants that can be satisfied beyond needs is an empirical question in itself. This is one of the reasons why we suggest to prioritise the provision of human needs in postgrowth and sustainability research and to think about subjective wellbeing within this framework (see the detailed discussions in Chap. 5 and Koch et al. 2017).

We regard Doyal and Gough's theory of human need as especially relevant for an academic understanding of global sustainable welfare due to the centrality of the notion of thresholds and limits that define different levels, and particularly the lowest level, at which human needs—understood as a "minimally decent life"—can be satisfied. Critical thresholds for the universal provision of human needs (and wants) or for a minimally decent life" are to be constantly (re-)defined in light of the advancement of scientific but also practical knowledge. For this auditing process, climate expertise, sustainability science as well as heterodox economics and social sciences could play important roles. These disciplines would, however, need to be integrated into a common steady state and sustainable welfare perspective. Whether it is possible to provide satisfaction of basic human needs on a global scale, and the extent to which more than basic needs (and wants) can be provided on a finite planet without overshooting ecological limits, are timely issues and questions for future research. Accordingly, we encourage projects that could start from Doyal and Gough's eleven intermediary needs and produce knowledge on how to best satisfy them in a world characterised by constraint and ecological crisis. A useful example is the ongoing debate on the ecological and socio-economic feasibility of nurturing all human beings on the basis of (local) organic farming (Badgley et al. 2007; Seufert et al. 2012; Tilman and Clark 2014; Koch et al. 2017). Though we can obviously not predict the outcome of such research and debate, we would follow Gough (2014: 378) in not ruling out the possibility that, due to climate change and ecological overshoot, less than the "optimal generalizable satisfaction of basic needs" can actually be achieved. Society would then need to find ways to democratically debate possible offsets between the satisfaction of the needs of current and future generations.

It follows from Doyal and Gough's parsimonious definition of basic human needs that these will be the same for future people as those of the present. And the same categories of universal satisfier characteristics apply: "Future people will have needs for affiliation, cognitive and emotional expression, understanding and critical thought" (Gough 2015: 13). This has repercussions for the structure of the present economy. In O'Neill's (2011: 33) words, "each generation needs to pass down the conditions for livelihood and good health, for social affiliation, for the development of capacities for practical reasoning, for engaging with the wider natural world and so on". Even though the present generation may be largely ignorant about the "detailed nature and quantum of need satisfiers that future peoples in future contexts will require" (Gough 2015: 14), Doyal and Gough (1991: 230–236) nevertheless clarify that all economic systems would need to be assessed according to their ability to produce enough appropriate need satisfiers. In an attempt to apply the protection-from-harm-of-future-generations principle to climate change and the environmental crisis, we will now outline in the most general terms what this may entail for the different regulatory scales of economy and society.

Towards a New Division of Labour Between Global, National and Local/Individual Levels

The planet's ecological footprint and its associated matter and energy throughput, particularly carbon emissions, would need to shrink significantly if the world's production and consumption patterns were to respect ecological limits. The currently richest countries would need to make the biggest contributions in a transition to a global SSE. One of the greatest obstacles for an adequate global governance system, highlighted in the literature on the international climate negotiations (Roberts and Parks 2006; Koch 2012), is enormous social inequalities (Piketty and Saez 2014) and the resulting absence of trust between "developing" and "developed" countries that constantly led to "solutions" based on the "law of the least ambitious program" (Gough 2013: 190). The outcome of the 2015 Conference of the Parties (COP21) summit in Paris is no exception to this rule. Though the agreement makes reference to an aspiration to keep global average temperature rises "well below" 2C—and even to make "efforts" towards 1.5C—it fails to specify what exactly is supposed to be done, "by whom, and to what degree" (Morgan 2016: 2), particularly in relation to the rich countries.

We are aware of the enormous difficulties that activists have experienced in establishing a more just global and multi-layered system of climate change governance and of the fact that there is no world government that could decide to turn far-reaching proposals such as a global SSE into reality. However, we believe it makes sense to use the remaining autonomy of the academic field to raise the issue of whether there are any indications that existing patterns of the global climate and environmental governance system may be used and/or adjusted in ways to help re-embed global production and consumption patterns into environmental limits in the future. First of all, and despite the fact that little has been achieved, the existence of such a governance structure, including the COP summits and the adoption of Sustainable Development Goals through the United Nations, can be interpreted as a confirmation of our previous point that global answers need to be sought for "wicked" environmental issues such as climate change. It is worthwhile remembering that the United Nations Framework Convention on Climate Change (UNFCCC) included the "precautionary principle", the "polluter pays" principle as well as the principle of "common but differentiated responsibilities" between rich and poor countries (Gough 2013). Furthermore, the establishment of the International Panel of Climate Change (IPCC) in 1988 to provide a clear, multidisciplinary and common scientific perspective on climate change, which assembles all reputable climate scientists and periodically produces consensus reports on the state of the problem, may be seen as a further step to "governance by scientific assessment" (Biermann 2011). Though the IPCC is limited to greenhouse gas emissions, its intentions go some way towards globally achieving caps for energy and matter throughput. Political science research could further analyse the prospects of this becoming reality.

We welcome the suggestion by "Earth System Governance" researchers to upgrade the UN Environment Programme to a UN specialised agency for environmental protection along the lines of the World Health Organisation. Similar to the Bretton Woods organisations, a "strong environmental organization with a sizable role in agenda-setting, norm-development, compliance management, science assessment, and building" (Biermann et al. 2012: 1306) would emerge. Also the proposal to replace the UN Commission on Sustainable Development by a new mechanism that stands much higher in the international institutional hierarchy—a "high-level UN Sustainable Development Council directly under the UN General Assembly" (Biermann et al. 2012: 1306)—appears to be a step in the right direction. However, whether, as these authors suggest, the primary members of this council should be the G20 countries is debatable. In this case, many of the countries that are historically responsible for the bulk of the problem would be given the largest amount of power to deal with it—at the expense of those countries that suffer the most and have the least possibilities to cope. It is worthwhile noting that it was the G77 group of countries, to which the "Earth Governance" theorists do not want to give primary member status in a

UN Sustainable Development Council, which has tabled, in relation to climate change and the individual level, the farthest-reaching proposal based on per capita pollution allowances.

Following a rather simple egalitarian logic, every global citizen would be allocated an equal entitlement to pollute the atmosphere according to specified individual emissions budgets in accordance with the global greenhouse gas reduction benchmarks identified by climate scientists (Grubb et al. 1999: 270; Davey and Douthwaite 2012). This would unavoidably call Western consumption patterns into question—including carbon-emission intensive practices such as individual flights and automobile use—since citizens of the developed countries have already used a disproportionate amount of their share of carbon and would therefore need to contract their carbon budget substantially. Citizens of developing nations, who have thus far emitted fewer greenhouse gases than their proportional share, could in contrast increase fossil fuel consumption for a certain period and eventually converge with the developed countries (Roberts and Parks 2006: 145). Rather than excluding most developing countries from environmental "management", the rich countries would need to clearly "signal their commitment" to restore trust though a "series of confidence-building measures" (Roberts and Parks 2006: 217). A further indispensable element for building trust at the international level would be the "greater stake" of developing countries in governance and decision-making of international financial institutions (Roberts and Parks 2006: 24).

The bio-physical terms and limits for the global matter and energy throughput defined at the global level would delineate the room for manoeuvre within which national and local economies and societies could evolve. In relation to the national level, Buch-Hansen (2014) has argued for an institutional perspective within postgrowth research, since the present institutional diversity of states is likely to affect national degrowth trajectories and the concrete shapes of national SSEs. If an introduction of SSEs were politically decided in the rich countries, these would most likely not follow one unified ideal type, but rather a range of different models, comprising a great diversity of institutional arrangements, actors and practices, affected by different economic, cultural and socio-technical contextual factors. Just as contemporary capitalist societies are diverse, so would SSEs take many different forms in different places as would the sustainable welfare institutions by means of which human needs would be satisfied. Theorists of institutional change

(Mahoney and Thelen 2010) argue that change rarely takes the form of an abrupt and clear-cut break with the past. More often change is gradual so that the institutional principles and practices of existing welfare arrangements would be preserved in some form and synthesised with steady-state and sustainable welfare principles (Koch and Buch-Hansen 2016). As a corollary, the institutions of national SSEs and their corresponding sustainable welfare systems would vary across space. This does not exclude the possibility of institutional learning processes from "best-practice" countries. Comparative research into wellbeing of existing countries relative to GDP per capita (Fritz and Koch 2014) suggests that there are better than average performing countries in each part of the world (for example, Switzerland for Europe, Costa Rica and Uruguay for Latin America) that could be singled out for future in-depth institutional analysis.

Chapter 3 addressed ongoing rescaling trends in welfare provision and social policy from national to local policy levels. This is of significance for the present discussion as a transition of regulatory power from traditional national welfare institutions to local levels could facilitate the satisfaction of human needs within globally determined ecological limits. Although not everything can or should be produced locally, several contributors to postgrowth debates have argued for the need to replace today's global capitalist system with economies based on principles of the cooperatives and social enterprise movement and oriented towards local production and consumption cycles (Dietz and O'Neill 2013). Yet it is not only from such academic output but also from emerging practices of new and different ways of locally producing and living (Howell 2012) such as ecovillages, transition towns and social enterprises that sustainable welfare systems can draw inspiration. Indeed, some of such voluntary grassroots initiatives have proven to be quite efficient in environmental terms even though they often face difficulties in sustaining themselves over time (Hildingsson and Koch 2016; Büchs et al. 2015).

More research into such local initiatives, their social composition and institutional embedding is necessary to support and facilitate the long-term achievement of environmental goals. We would expect the chances of achieving long-term success to increase where (local) governments and governance networks support voluntary and civic bottom-up initiatives (Soper 2016). Further theoretical examination and empirical evidence should be produced on how elements of the "good life" are conceived locally. We may already find embryonic forms of alternative visions and practices in craft and art movements, the service economy, socio-ecological enterprises and collaborative consumption initiatives. Corresponding research has the potential of contributing towards creating a common vision of a social order where individual lifestyles of personal fulfilment and enjoyment are complemented by environmentally sound and socially just production and consumption methods, and associated modes of governance at local, national, regional and global levels (Mont and Koch 2016). "Slower" and local lifestyles with more free time, for example, would then not be seen as the end of occidental culture but as sources of individual and collective wellbeing. At the same time, this would be a mechanism to sever the link between resource-intensive economic growth and societal progress—a severance that we regard as a necessary precondition for overcoming the domination of the shopping mall culture over alternative and sustainable definitions of well-being and quality of life.

Eco-Social Policies for Postgrowth

Steady-state economics and sustainable welfare are oriented towards the satisfaction of human needs within ecological limits, from intergenerational and global (intra-generational) perspectives. This is reflected in Gough's "policy auditing" (2015) approach, according to which existing economic, social and environmental policies as well as material welfare standards would need to be reviewed according to the criterion of generalisability. Beyond basic human needs, material welfare and wellbeing would be regarded as secondary to environmental sustainability (Koch and Mont 2016). Production and consumption patterns would need to be organised in ways that the global matter and energy throughput and associated bio-physical flows do not exceed levels identified by sustainability sciences. Accordingly, economic growth as a policy goal would need to be deprioritised relative to the satisfaction of human needs within ecological limits. Postgrowth policy approaches have remained at fairly abstract levels to date, mostly failing to discuss concrete policy proposals, let alone their synergy potentials in a coherent transition strategy. Exceptions to this rule are Daly (2013), Dietz and O'Neill (2013), and Davey and Douthwaite (2012) (amongst others) who propose policies ranging from cap-and-trade/share systems for basic resources, ecological tax reforms, re-regulations of working hours and international commerce, to maximum limits on income and wealth and minimum income guarantees to reduce social inequality which we can use as point of departure. A quick look at this list already suggests its implementation would require a new mix of property forms including communal, state and individual property and a new division of labour between market, state and "commons", where markets would play a much lesser role than currently. Or, in Marxian terms, global capitalism would need to be "overcome" in the sense that the dominance of monetary growth or exchange value over bio-physical parameters or use value would be reversed (Koch 2015). Hence, most of the following policy proposals have the potential of becoming "real utopias" to transcend contemporary capitalism in the sense of E.O. Wright (2013).

In rich countries, the state's traditional and growth-oriented steering role in economic and socio-ecological policy making (Chap. 3) would need to change substantially if economic growth were to be deprioritised and bio-physical parameters as well as the satisfaction of basic human needs prioritised. The provision of sufficient needs satisfiers for all people now and in the future presupposes the redistribution of economic resources and the definition and implementation of limits for economic and social inequality which would need to be much narrower than in the current capitalist growth economy. To this end, some sort of "distributist institutions" (Koch and Buch-Hansen 2016) would be necessary to steer a range of "eco-social policies" in the transition towards ecological and social sustainability. Generally, these would need to address the "double injustice" (Walker 2012) that the poorest household groups who are least responsible for environmental damages such as climate change are in the worst position to cope and to afford mitigation and adaption (Büchs et al. 2011). Ecological investment into retrofitting houses, for example, has only a chance of being acceptable to the electorate if it is accompanied by countervailing social policies (Gough 2011) that assist homeowners in affording ecologically useful measures. More generally, through targeted policies, states can help bring about a redistribution of work, wealth and pollution rights and can stimulate alternative ways of consumption. Focusing on the rich countries, we review postgrowth reform ideas concerning the policy areas of macroeconomic steering, inequality/redistribution, minimum and maximum incomes, carbon rationing, consumption, working time reduction and work–life balance as well as population/migration. Other important proposals and approaches which we cannot review due to space constraints include the role of commons; the cooperative economy and communal forms of living; as well as alternative monetary systems and local currencies (Bollier and Helfrich 2012; Chatterton 2016; Davey and Douthwaite 2012).

In an attempt to map out economies in which GDP growth is sidelined and where bio-physical indicators, stability and resilience are in focus, Daly and Farley (2011: 55) suggest two main principles of macroeconomic reforms that respect ecological limits; firstly, the rate of extraction of non-renewable resources should not exceed the rate of creation of renewable substitutes, and secondly, waste emissions should not exceed the environment's capacity to absorb them. Daly and Farley but also Jackson argue that achieving these goals cannot be left to the market but requires an active state to set a collective limit on aggregate throughput to keep it within the absorptive and regenerative capacities of the ecosystem. They are also in favour of a re-regulation of the international political economy away from free trade, free capital mobility and unregulated financial markets, and promote local economic circles instead. Jackson (2009: 104) advocates an increase in public control of the money supply to provide greater protection against consumer debt. He also demands public sector jobs in building and maintaining public assets, investments in renewable energy, public transport infrastructure and public spaces, strengthening community-based sustainability initiatives and especially the retrofitting of the existing building stock with energy- and carbon-saving measures. Finally, all postgrowth authors are in favour of investment into ecological transitions in developing countries, renewable energy, resource efficiency, low-carbon infrastructures, and the protection of habitats and biodiversity. At company level, both Daly and Farley (2011) and Wilkinson and Pickett (2009) demand public intervention in the existing property structure and, in particular, a broadening of capital ownership to regulate workplace-based structures of inequality and rank-ordered hierarchies.

There is agreement amongst postgrowthers that the distribution of wealth and income both within and across countries and in an intergenerational perspective is crucial for the reduction of carbon emissions (Daly and Farley 2011: 441; Koch 2012: 178–193). Generally, Daly and Farley (2011: 442) propose that government redistribution policies should respect what people have earned through their own efforts, but people should "not be able to capture for themselves values created by nature, by society, or by the work of others. And they should pay a fair price for what they receive from others, including the services provided by government, and for the costs they impose on others". To achieve redistribution and to enhance ecological sustainability, most postgrowth authors also argue for an ecological tax reform. Jackson (2009: 106) outlines its general direction by shifting the burden of taxation from "economic goods (e.g. incomes) to ecological bads (e.g. pollution)". If the tax base were linked to the throughput of finite resources, external costs, which private enterprises enjoy as "free gifts" from nature to date, would be internalised and considered in their cost calculations. However, Daly (1977: 63) prefers the definition of depletion quotas to pollution taxes, since the latter would increase competition within the recycling industry. Furthermore, eco-taxes are often regressive in distributional terms, so would have to be accompanied by redistributive policies financed through their revenues or replaced by cap-and-share or traded pollution quotas which have progressive distributional effects (Büchs et al. 2014; Davey and Douthwaite 2012). Some approaches also suggest to complement (regressive) depletion and/or pollution taxes with progressive income and inheritance tax reforms. For instance, Daly and Farley (2011: 444) advocate a "highly progressive income tax that asymptotically approaches 100%, more direct limits on how much someone can earn, or relative limits that establish a legal ratio between the highest and lowest income allowed" and "very high inheritance taxes" since much of the accumulated wealth is inherited. Any ecological tax reform would also need to be combined with a shift in the tax burden from taxes on labour to taxes on activities causing environmental damage, high-carbon luxuries, profits and rents and possibly land (Davey and Douthwaite 2012).

We agree with most postgrowthers that the definition of both minimum and maximum limits on income and wealth is critical for any steady-state transition. After reaching the maximum income, people would be incentivised to "devote their further energies to noneconomic pursuits" so that confiscatory revenues would be rather small. The opportunities thus forgone by the wealthy would be made available to the "not-so wealthy, who would still be paying taxes on their increased earnings. The effect on incentive would be negative at the top but positive at lower levels, leading to a broader participation in running the economy" (Daly 1977: 56). Varying across authors, a minimum or basic income would be co-financed from general revenues, an increasingly progressive income tax, eco-taxes and/or from depletion and emissions certificate auctions. Andersson (2009: 3), who reviewed various attempts of linking postgrowth approaches and basic income schemes, suggests equivalence between basic income financed by green taxes and the distribution of equal and transferable rights to use scarce environmental resources and to emit a given quantity of greenhouse gases. In line with the hypothesis that it will ultimately be necessary to limit transnational and global inequalities in wealth and income in order to reach an Earth-wide steady state, Andersson (2009: 6) proposes the successive generalisation of an unconditional basic income from the already rich countries to a global scheme.

There is consensus amongst postgrowth authors on the necessity of identifying clear resource and emission caps according to climate science expertise and on the establishment of reduction targets under those caps. There is further agreement on the application and generalisation of "contraction and convergence" and "cap-and-share" models for climate-related emissions at equal per capita allowances (Jackson 2009: 106; Davey and Douthwaite 2012), leading to the eventual convergence of equal per capita emissions across the planet. The British Sustainable Development Commission (2007: 7), for example, advocated the introduction of a measurement of individual carbon footprints as a central element of the measurement of environmental wellbeing. This indicator would need to reflect "not only the direct emissions associated with consumption in the UK, but also the emissions "embedded" in imported goods and services". The consensus is that if policies to cut emissions were to be seen as fair, richer persons and countries, which on average contribute much more to climate change than poorer persons and countries, would be affected the most (Büchs et al. 2011; 2014). Daly suggests the introduction of pollution allowances and tradable rights, for example, for carbon emissions. However, empirical evidence on the effectiveness of real-existing market-based instruments such as the European Union Emissions Trading Scheme (Koch 2014) or Personal Allowances and Trading schemes (Hildingsson and Koch 2016) does not give much reason to cheer. These are severely hampered by the fact that the reduction of carbon emissions is intended to be the by-product of the profit/market logic in general and the search for new investment opportunities for finance capital in particular.

The postgrowth view of Western consumption rates is that they would need to decrease disproportionately so that citizens of other countries could enjoy an improvement in their material standard of living. While consumption is generally seen as critical to human development as long as it "enlarges the capabilities of people without adversely affecting the well-being of others" (The Royal Society 2012: 47), there is agreement with Daly and Farley's (2011: 442) argument that on a finite planet the present generation should develop a "sense of obligation toward future generations" that is seen as being entitled to having the same opportunities for development as the present. Conspicuous consumption is then a "negative externality", and people should pay for the negative impacts this imposes upon others. Policy proposals about the most effective ways of reducing such consumption and the accompanying carbon emissions are not very detailed as yet, but could consist of progressive consumption taxes (Daly and Farley 2011: 444) or a global cap-and-share scheme which would limit the fossil fuel input to the economy and hence fossil fuel-based production and consumption (Davey and Douthwaite 2012). Kasser (2009: 178) suggests a threefold strategy involving the decrease of the extent to which people are exposed to lifestyle models of conspicuous consumption, for example, by banning advertisements aimed at children; the support of people's resilience, for example, by teaching individuals how to decode advertisement messages; and helping people to act in accordance with "intrinsic" goals, for example, by encouraging ethical consumption. Finally, on top of financial incentives, governments can encourage certain ways of consumption (e.g. vegetarian diets, local holidays, use of public transport and cycling) and make others less attractive (e.g. meat consumption, holidaying in distant locations, car and plane use). The success of such state engagement may be facilitated by a growing dissatisfaction of the public with the consumerist lifestyle (Soper 2016).

If physical indicators of throughput and GDP as a whole are reduced and labour productivity does not decline, growing unemployment would be the result. Postgrowth theorists have therefore started to debate the relations between postgrowth, remuneration, employment and work (Martínez-Alier et al. 2010: 1746). Moving towards an SEE would entail a significant cut in the percentage of time spent in paid work in order to reduce unemployment and distribute working time more evenly across the population, break the circle of working to earn to consume, and to enable a better work–life balance as well as time for currently unpaid activities such as childcare and personal care or engagement in local voluntary activities (Koch and Fritz 2013). In most approaches, the welfare state plays a crucial role in this redistribution (Gough and Meadowcroft 2011: 500). Reducing the working week is, for example, at the heart of Victor's resilience scenario for the Canadian economy.

Victor (2010: 371) suggests that employment can be spread more evenly amongst the workforce allowing the "benefits of greater productivity" to be "directed towards more leisure time, rather than increasing GDP", thanks to shorter working hours as key ingredient. From a more general theoretical perspective, such readjustment of employment, work and other activities presupposes placing "both on a more equal footing, rather than seeing 'work' as signifying a lack, or a less valuable human activity than 'employment'" (Barry 2012: 139). This, again, calls for a "more expansive conceptualisation of the economy in which all work, all economic activity, all resource and energy use is included" (Barry 2012: 139). In a recent comparative overview of working time reduction policies, which were frequent in OECD countries prior to the 1990s, Mont (2016) concludes that these were especially successful in improving the work/life balance, while their effect on generating employment was weaker. However, she stresses that none of these policies were devised to meet the triple goal of increased employment, wellbeing and a simultaneous reduction of environmental impacts. Models such as those provided by Victor (2008) and Malmaeus (2011) indicate that achieving this triple goal would require a combination of measures, including an environmental tax reform and/or cap-and-share schemes, basic income as a measure to counter the adverse effects of growing productivity and labour replacements.

Finally, many postgrowthers argue that a global SSE would ultimately be predicated on relatively stable population levels, since—all other things being equal—more people generate more greenhouse gas emissions and use up more finite resources than fewer people. This goal raises the issue of appropriate population sizes and of suitable ways of achieving this. Originally, Daly (1977: 57) advocated a scheme of "transferable birth licences", according to which every woman would receive an amount of reproduction licenses that corresponds to replacement fertility. However, such a "population stabilisation institution" turned out not only to be very controversial in the debate given its authoritarian undertones, but is also of limited applicability to European and OECD countries where birth rates have been shrinking considerably over the last decades. Tax breaks for families with fewer children in combination with more generous immigration policies would, in our view, be the short-term way forward. In the longer term, we would assume that an introduction of minimum and maximum income levels across the globe and a corresponding decrease in international inequality would serve as a policy nudge towards smaller family sizes as old-age pensioners, for example, would become less dependent on informal exchanges within families.

Conclusion

This chapter has outlined what we regard as the most general possible features of a global SSE and an according "sustainable welfare" system. We argued for an extension of the steady-state concept to the global level where critical thresholds for energy and matter throughput would need to be defined. These would at the same time determine the bio-physical limits within which national and local economies could evolve. Indeed, if global thresholds were to be met, local levels would need to play a much greater role in production and consumption than currently, and it will belong to the crucial future tasks to further develop the links between global and local regulatory levels. Given ecological overshoot and short time frames for climate action, we regard it as unlikely that "wellbeing" for all and for future generations can for the time being mean much more than the satisfaction of basic human needs. We have therefore highlighted the significance of the concept of human needs for postgrowth research and any sustainable welfare system both here and in Chap. 5. Our final review of existing "eco-social" policy proposals, which may facilitate the transition towards an SEE and sustainable welfare, has been, on the one hand, encouraging, since there actually already exists a list of more or less developed policy suggestions to which activists may turn. On the other hand, this review has confirmed the previous observation that these partially far-reaching policy proposals are still mainly studied "within separate silos" (Gough 2011: 59). This may not be surprising given the fact that the theoretical approaches upon which these proposals are built are likewise diverse and in need of integration. However, as in the case of the different postgrowth theories, there appears to be sufficient common ground for combining, complementing and unifying the as yet fragmented policy proposals and for formulating a coherent strategy for the economic, political and ecological restructuring of the advanced capitalist countries.

Note

1. Steinberger and Roberts (2010) suggest that this might be possible as already achieved by countries with moderate resource use.

References

Andersson, J.O. 2009. Basic Income from an Ecological Perspective. *Basic Income Studies* 4 (2): 1–8.

Badgley, C., J. Moghtader, E. Quintero, E. Zakem, M. Jahi Chapelli, K. Aviles-Vazquez, A. Samulon, and I. Perfecto. 2007. Organic Agriculture and the Global Food Supply. *Renewable Agriculture and Food Systems* 22 (2): 86–108.

Barry, J. 2012. Towards a Political Economy of Sustainability. In *Climate Change and the Crisis of Capitalism*, ed. M. Pelling, D. Manuel-Navarrete, and M. Redclifft, 129–141. London: Routledge.

Biermann, F. 2011. New Actors and Mechanisms of Global Governance. In *The Oxford Handbook of Climate Change and Society*, ed. R. Dryzek, B. Norgaard, and D. Schlosberg, 685–695. Oxford: Oxford University Press.

Biermann, F., K. Abbot, S. Andresen, K. Bäckstrand, S. Bernstein, M.M. Betsill, H. Bulkeley, B. Cashore, J. Clapp, C. Folke, A. Gupta, J. Gupta, P.M. Haas, A. Jordan, N. Kanie, T. Kluvankova-Oravska, L. Lebel, D. Liverman, J. Meadowcroft, R.B. Mitchell, P. Newell, S. Oberthür, L. Olsson, P. Pattberg, R. Sanchez-Rodriguez, H. Schroeder, A. Underdal, S. Camargo-Vieira, C. Vogel, O.R. Young, A. Brock, and R. Zondervan. 2012. Navigating the Anthropocene: Improving Earth System Governance. *Science* 335: 1306–1307.

Bollier, D., and S. Helfrich. 2012. *The Wealth of the Commons: A World Beyond Market and State*. Amherst, MA.: Levellers Press.

Büchs, M., Duwe, S., and Bardsley, N. 2011. Who Bears the Brunt? Distributional Effects of Climate Change Mitigation Policies. *Critical Social Policy* 31 (2): 285–307.

Büchs, M., N. Bardsley, and S. Schnepf. 2014. Unequal Emissions—Unequal Policy Impacts: How Do Different Areas of CO_2 Emissions Compare? In *International Handbook on Social Policy and the Environment*, ed. T. Fitzpatrick, 62–92. Cheltenham, UK: Edward Elgar.

Büchs, M., C. Saunders, R. Wallbridge, G. Smith, and N. Bardsley. 2015. Identifying and Explaining Framing Strategies of Low Carbon Lifestyle Movement Organisations. *Global Environmental Change* 35: 307–315.

Chatterton, P. 2016. Building Transitions to Post-Capitalist Urban Commons. *Transactions of the Institute of British Geographers* 41 (4): 403–415.

Daly, H. 1972. In Defense of a Steady-State Economy. *American Journal of Agricultural Economy* 54 (5): 945–954.

Daly, H. 1977. *Steady State Economics*. San Francisco, CA: W.H. Freeman.

Daly, H. 2013. *'Top 10 Policies for a Steady-State Economy', Centre for the Advancement of the Steady State Economy*. http://steadystate.org/top-10-policies-for-a-steady-state-economy/.

Daly, H., and J. Farley. 2011. *Ecological Economics. Principles and Applications*, 3rd ed. Washington: Island Press.

Davey, B., and R.J. Douthwaite. 2012. *Sharing for Survival: Restoring the Climate, the Commons and Society*. Dublin: FEASTA.

Dietz, R., and D. O'Neill. 2013. *Enough is Enough. Building a Sustainable Economy in a World of Finite Resources*. London: Earthscan/Routledge.

Doyal, L., and I. Gough. 1991. *A Theory of Human Need*. Basingstoke: Palgrave Macmillan.

Farley, J. 2013. Steady State Economics. In *Degrowth: A Vocabulary for a New Era*, ed. G. D'Alisa, F. Demaria, and G. Kallis, 49–52. London: Routledge.

Fritz, M., and M. Koch. 2014. Potentials for Prosperity Without Growth: Ecological Sustainability, Social Inclusion and the Quality of Life in 38 Countries. *Ecological Economics* 108: 191–199.

Fritz, M., and M. Koch. 2016. Economic Development and Prosperity Patterns Around the World: Structural Challenges for a Global Steady-State Economy. *Global Environmental Change* 38: 41–48.

Gough, I. 2011. *Climate Change and Public Policy Futures*. London: British Academy.

Gough, I. 2013. Climate Change, Social Policy, and Global Governance. *Journal of International and Comparative Social Policy* 29 (3): 185–203.

Gough, I. 2014. Lists and Thresholds: Comparing the Doyal-Gough Theory of Human Need with Nussbaum's Capabilities Approach. In *Capabilities, Gender, Equality. Towards Fundamental Entitlements*, eds. F. Comin, and M. Nussbaum, 357–381. Cambridge: Cambridge University Press.

Gough, I. 2015. Climate Change and Sustainable Welfare: The Centrality of Human Needs. *Cambridge Journal of Economics* 39 (5): 1191–1214.

Grubb, M., C. Vrolijk, and D. Brack. 1999. *The Kyoto Protocol: A Guide and Assessment*. London: Royal Institute of International Affairs.

Hildingsson, R., and M. Koch. 2016. Market Solutions to Climate Change: Examples of Personal Carbon-Trading and Carbon-Rationing. In *Sustainability and the Political Economy of Welfare*, ed. M. Koch, and O. Mont, 109–124. London: Routledge.

Howell, R.A. 2012. Living with a Carbon Allowance: The Experiences of Carbon Rationing Action Groups and Implications for Policy. *Energy Policy* 41: 250–258.

IPCC. 2014. *Climate Change 2014: Synthesis Report—Summary for Policy Makers. Contribution of Working Groups I, II and III to the Fifth Assessment Report of the Intergovernmental Panel on Climate Change*. Geneva: Intergovernmental Panel on Climate Change.

Jackson, T. 2009. *Prosperity without Growth? The Transition to a Sustainable Society*. London: Sustainable Development Commission.

Jackson, T., and P. Victor. 2011. Productivity and Work in the New Economy—Some Theoretical Reflections and Empirical Tests. *Environmental Innovations and Societal Transitions* 1 (1): 101–108.

Kasser, T. 2009. Psychological need Satisfaction, Personal Wellbeing, and Ecological Sustainability. *Ecopsychology* 1 (4): 175–180.

Koch, M. 2012. *Capitalism and Climate Change. Theoretical Discussion, Historical Development and Policy Responses.* Basingstoke: Palgrave.

Koch, M. 2014. Climate Change, Carbon Trading and Societal Self-Defence. *Real-world Economics Review* 67: 52–66.

Koch, M. 2015. Climate Change, Capitalism and Degrowth Trajectories to a Global Steady-State Economy. *International Critical Thought* 5 (4): 439–452.

Koch, M., and H. Buch-Hansen. 2016. Human Needs, Steady-State Economics and Sustainable Welfare. In *Sustainability and the Political Economy of Welfare*, ed. M. Koch, and O. Mont, 29–43. London: Routledge.

Koch, M., and M. Fritz. 2013. Non-Standard Employment: Concept, Empirical Results and Policy Implications. In *Non-Standard Employment in Europe: Paradigms, Prevalence and Policy Implications*, ed. M. Koch, and M. Fritz. Basingstoke: Palgrave Macmillan.

Koch, M., and O. Mont (eds.). 2016. *Sustainability and the Political Economy of Welfare.* London: Routledge.

Koch, M., H. Buch-Hansen, and M. Fritz. 2017. Shifting Priorities in Degrowth Research: An Argument for the Centrality of Human Needs. *Ecological Economics* 138: 74–81.

Mahoney, J., and K. Thelen (eds.). 2010. *Explaining Institutional Change. Ambiguity, Agency and Power.* Cambridge: Cambridge University Press.

Malmaeus, M. 2011. *Ekonomi Utan Tillväxt. Ett Svenskt Perspektiv.* Stockholm: Cogito.

Martinez-Alier, J., U. Pascual, F.D. Vivien, and E. Zaccai. 2010. Sustainable De-Growth: Mapping the Context, Criticism and Future Prospects of an Emergent Paradigm. *Ecological Economics* 69: 1741–1747.

Milanovic, B. 2005. *Worlds Apart. Measuring International and Global Inequality.* New Jersey: Princeton University Press.

Mont, O. 2016. The Changing Landscape of Worktime Reduction. The Past and the Future. In *Sustainability and the Political Economy of Welfare*, eds. M. Koch, and O. Mont, 125–140. London: Routledge.

Morgan, J. 2016. Paris COP 21: Power that Speaks the Truth? *Globalizations* 13 (6): 1–9.

O'Neill, J. 2011. The Overshadowing of Needs. In *Sustainable Development*, ed. F. Rauschmayer, I. Omann, and J. Frühmann, 25–43. London: Routledge.

O'Neill, D. 2012. Measuring Progress in the Degrowth Transition to a Steady-State Economy. *Ecological Economics* 84: 221–231.

O'Neill, D. 2015. The Proximity of Nations to a Socially Sustainable Steady-State Economy. *Journal of Cleaner Production* 108: 1213–1231.

Piketty, T., and E. Saez. 2014. Inequality in the Long Run. *Science* 344 (6186): 838–843.

Roberts, J.T., and B.C. Parks. 2006. *A Climate of Injustice. Global Inequality, North-South Politics, and Climate Policy*. Cambridge: The MIT Press.

Seufert, V., N. Ramankutty, and J.A. Foley. 2012. Comparing the Yields of Organic and Conventional Agriculture. *Nature* 485: 229–232.

Soper, K. 2016. The Interaction of Policy and Experience: An "Alternative Hedonist" Optic. In *Sustainability and the Political Economy of Welfare*, ed. M. Koch, and O. Mont, 186–200. London: Routledge.

Steinberger, J.K., and J.T. Roberts. 2010. From Constraint to Sufficiency: The Decoupling of Energy and Carbon from Human Needs, 1975–2005. *Ecological Economics* 70 (2): 425–433.

Sustainable Development Commission. 2007. Living Well—Within Limits. SDS Discussion Document on Wellbeing Indicators for Sustainable Development. London.

The Royal Society. 2012. People and the Planet. The Royal Society Science Policy Centre report 01/12. London. https://royalsociety.org/topics-policy/projects/people-planet/report/.

Tilman, D., and M. Clark. 2014. Global Diets Link Environmental Sustainability and Human Health. *Nature* 515: 518–522.

Victor, P.A. 2008. *Managing Without Growth: Slower by Design, Not Disaster*. Cheltenham: Edward Elgar.

Victor, P.A. 2010. Questioning Economic Growth. *Nature* 468: 370–371.

Walker, G. 2012. *Environmental Justice: Concepts, Evidence and Politics*. London: Routledge.

Wilkinson, R.G. and K.E. Pickett. 2009. The Spirit Level. *Why More Equal Societies Always Do Better*. London: Allen Lane.

Wright, E.O. 2013. Transforming Capitalism Through Real Utopias. *American Sociological Review* 78 (1): 1–25.

CHAPTER 8

Conclusions

Abstract This book combines political economy and social practice perspectives to highlight the challenges of achieving wellbeing goals under postgrowth. Our review of assumptions about wellbeing in postgrowth debates first highlights potential problems with theories of adaptive preferences. While preferences may easily adapt upwards when living standards improve, the concept of loss aversion and evidence on wellbeing in times of economic crisis suggest that they do not adapt equally well downwards. Second, social practice theory helps us understand the enormous challenges which will be involved in decoupling the generation of wellbeing from its current embeddedness in growth-based market capitalism and a range of other closely linked structures. Conceptually, the debate should focus on basic human needs as is it compatible with postgrowth premises. Finally, institutional contexts, especially those that promote greater social equality, will be crucial for supporting wellbeing under postgrowth.

Keywords Postgrowth · Wellbeing · Social practices · Political economy · Human needs

The postgrowth literature argues that economic growth is not sustainable from environmental and social perspectives. It calls for a shrinking of material throughput of the economy in developed Western countries to then enter a steady state at a sustainable scale. This book agrees with

M. Büchs and M. Koch, *Postgrowth and Wellbeing*,
DOI 10.1007/978-3-319-59903-8_8

this analysis but seeks to advance the discussion about possible wellbeing implications of postgrowth which has so far been too polarised and uncritical. However, we have demonstrated in this book that the relationships between economic growth and decline on the one hand, and wellbeing on the other, are more complex than often portrayed in this debate and require further analysis and research. An important contribution that this book makes is to draw on insights from social practice theory and political economy to examine the ways in which wellbeing is currently embedded in an institutionalised growth paradigm in society. This analysis and future research that may follow from it are important to enable policy makers, practitioners and researchers to identify the conditions that are likely to support wellbeing in the context of postgrowth.

Postgrowth proponents support the idea that wellbeing can be maintained or even improved in this new context with three key arguments. The first emphasises that we need to take a long-term perspective for thinking about the relationship between growth and wellbeing. Climate change is already affecting people's wellbeing in many places across the globe. If economic growth continues, its environmental impacts, especially from global warming, may be so extensive that human wellbeing is going to be severely undermined. From this perspective, it becomes a moral imperative to move away from growth-based societies now to preserve the rights of future generations to be well. The second argument refers to evidence that GDP growth is not clearly related to increases of subjective and objective wellbeing, especially if this is examined over longer time periods. The third argument critiques currently dominant definitions of wellbeing which equate wellbeing with income and material consumption, or utilitarian conceptions of happiness. Instead, postgrowth supporters tend to refer to eudemonic conceptions of wellbeing and a fulfilled, "good life", or notions of universal human needs. They argue that achievement of purpose and fulfilment in life and meeting human needs mainly rely on supportive human relationships and meaningful activities, both of which are thought to be facilitated by slower, more cooperative and less stressful lives beyond a growth economy (Chap. 5).

These are all very valid points, but in this book, we highlight several issues that require further attention. First, evidence that does not show a correlation between rising GDP and subjective wellbeing over time may be related to measurement issues. Unlike GDP, happiness and life satisfaction a measured on bounded scales which remain static over time and are hence of limited value of reflecting changes of actual levels of wellbeing. Furthermore, cross-sectional analysis within and across countries still shows that subjective and objective measures of wellbeing tend to be higher with higher levels of income. The flattening of subjective wellbeing or life expectancy curves at higher levels of income might again be related to issues of bounded scales and the fact that the distribution of income often increases exponentially at the higher end of the scale. Once logarithmic income measures are used, there is often a nearly linear relationship between income and measures of wellbeing (Deaton 2008).

Second, the relationship between income and wellbeing might differ if we compare phases of economic growth with phases of economic contraction. The concept of loss aversion (Tversky and Kahneman 1991) suggests that while people seem to adapt their expectations relatively fast to increased levels of living standards, they respond more strongly—and negatively—to welfare losses, especially if they occur rapidly. This idea is supported by empirical evidence which shows that both subjective and objective wellbeing (our focus here was on life expectancy) tend to decline during phases of economic contraction. The argument that wellbeing also recovers again with subsequent growth, as for instance shown in a paper by Easterlin (2010), cannot be applied to postgrowth because the economy would continue to shrink for an extended period of time before entering a phase of zero growth. Third, postgrowth supporters often highlight the importance of greater social equality to support wellbeing in phases of economic contraction. This is an important point as previous research has demonstrated the detrimental effects that social inequality can have on various wellbeing outcomes at the aggregate level (Wilkinson and Pickett 2009). While the argument has been made that, under certain circumstances, it is possible to prevent inequality from increasing during phases of economic contraction (Jackson and Victor 2016), other economists are more pessimistic on this question. Levels of inequality are likely to depend on the level of redistributive policies and market regulation which relies on political consensus. Generally, Chap. 5 has indicated that the evidence on the wellbeing impacts of economic contraction remains contradictory and thus suggests that social and policy contexts are likely to play an important role in mediating outcomes. Further research is required to identify the types of contexts that support wellbeing in phases of economic contraction.

Our analysis has also demonstrated that the current postgrowth debate has so far not been very coherent (and sometimes not even sufficiently explicit) regarding the concepts of wellbeing that are applied. Notions of (narrow) conceptions of subjective wellbeing coexist in the debate with those of eudemonic wellbeing as well as a variety of "alternative" objective concepts and measures of welfare or wellbeing. In this book, we support recent attempts to connect concepts of basic human needs with debates on postgrowth (Koch et al. 2017). These two approaches are very complementary because human needs are satiable (unlike wants) and non-substitutable (e.g. money cannot buy supportive personal relationships). The notion of basic human needs therefore lends itself to the identification of levels of material living standards that are both sufficient to satisfy human needs and remain within sustainable scales of material throughput to the economy. In Chap. 5, we have argued that the ethical obligation to preserve conditions for the wellbeing of future generations should be connected to the framework of basic human needs at the highest level which defines ultimate goals of human needs. In Doyal and Gough's (1991) framework, the ultimate goal of human needs is the avoidance of serious harm, which is defined as the "fundamental disablement in the pursuit of one's vision of the good, whatever that vision is" (ibid.: 50). We argue here that collective decision-making processes which define and implement this "vision of the good" need to take concern for the wellbeing of future generations into account to be compatible with postgrowth frameworks.

Furthermore, we have argued that further work is required to conceptualise the interrelationships between objective and subjective wellbeing. We wholly agree with critiques of purely subjective approaches to wellbeing and support the emerging greater emphasis on objective wellbeing approaches in the postgrowth debate. However, it remains important to consider the role that (necessarily subjective) understandings of wellbeing in society play for the acceptance of a shrinking economy and decreasing levels of material living standards—and the alternative visions of wellbeing that might be related to it. An academic agreement on maximum levels of material living standards for the fulfillment of basic needs and achievement of a satisfying life remains insufficient for the pursuit of postgrowth if the majority of the population disagrees and remains attached to material aspirations.

Finally, this book seeks to make a contribution to conceptualising wellbeing by linking it more closely to insights from social practices theory and political economy. Combining these approaches enables us to argue that the generation of wellbeing is currently closely embedded in dominant socio-eco-technological formations that rely on and are driven by economic growth through market capitalism. We also argue that the extent of the challenge of transitioning away from this formation should not be underestimated due to the close links that have become established between the constituent fields in this formation. In Chap. 2, we demonstrated how tightly coupled the growth paradigm and market capitalism are. The current form of capitalism inherently relies on growth. Moving to postgrowth hence requires the establishment of a different economic system which is not dependent on the generation of profit and continuous accumulation of capital—how this system may be called and to what extent it can build on existing institutions is a secondary question, but that its core logic, and related mechanisms and practices will qualitatively differ from the current form of market capitalism seems obvious.

Chapter 3 discussed the links between the welfare state and wellbeing—and the ways in which this may be affected by the contradictions that characterise the welfare state in market capitalism (Offe 1984): on the one hand, the welfare state limits market forces and the commodification of labour and thus contributes in manifold ways to people's wellbeing, including through the education, health, social care, old age, social security and minimum income systems that it provides. On the other hand, the welfare state cannot simply be regarded as a force that opposes the market. Quite the opposite, it evolved in tandem with the development of market capitalism and in many ways legitimised and stabilised this regime by supporting education, health and regeneration of the workforce. Therefore, current welfare state regimes also "lock" the generation of wellbeing into the functioning of growth-based capitalism and hence into a regime that is unsustainable and will undermine wellbeing in the long run. Paradoxically, while the welfare state currently contributes to the generation of wellbeing, partly because it limits market mechanisms and possibly to some extent growth, it also couples these processes to the unsustainable growth paradigm and supports growth in the long term, thus crowing out alternative wellbeing conceptions and practices.

In Chap. 6, we connected this discussion to social practice theory which emphasises the co-constitutive role of macro- and micro-dimensions of society—or what is often called structure and agency. We highlighted several implications that this perspective has for discussing wellbeing in the context of postgrowth. First, this perspective helps to demonstrate the structural properties that the growth paradigm assumes as it becomes embedded in various kinds of coupled socio-technical structures. Not only that, but it also becomes something that is perceived as "natural" by actors and thus unquestioningly reproduced on a daily basis as workers and welfare benefit recipients (have to) "buy into" the logic that labour market participation is to their own advantage and internalise aspirations of career progression and increasing incomes and consumption (also see Chap. 2). Second, social practice theory has emphasised the ways in which practices simultaneously create and are "held together" by close links between various social fields, and them and technological and environmental structures. This also happens because the constitution of these structural features of societies implies a concentration of material and immaterial resources, including power, in these socially dominant formations. While this distributional structure of socio-eco-technological formations requires continuous reproduction through practices that actors engage in, and is hence constantly recalibrated in the process, it is not easily altered more radically and rapidly. Radically different practices that do not fit into these established formations do not have access to the resources and power required to be "scaled up" and are marginalised in their encounters with dominant practices. Third, these insights are directly relevant to our discussion of wellbeing in the context of postgrowth. Essentially, wellbeing outcomes are generated through social practices which, as we have discussed above, are firmly embedded in and continually reproduce the currently dominant system of growth-dependent market capitalism and its links to a range of other socio-eco-technological systems.

An important question that emerges from this is in which ways the rapid and radical social change that will be required to move to a postgrowth society might impact on wellbeing. While the concept of agency, and with that the capacity for change, is fundamental to the social practices approach, it is also important to acknowledge the possibility that some dimensions of social structures might play a positive role for people's wellbeing because they provide people with identities, meanings, trust in relationships and institutions, security—all of which seem to be vitally important for leading fulfilled lives and for meeting basic human needs. Here we encounter again the difficulties that the links between subjective and objective wellbeing present us with: the previous statement may be true even though current social processes generate considerably unequal wellbeing outcomes for different groups in society and are unlikely to be in the in the "objective" interest for people's wellbeing in the long term. In other words, existing and relatively stable socio-eco-technological formations can provide people with identities, meanings and perceived security and thus contribute to their sense of wellbeing, even though not in a way that is "objectively" optimal. Does from this follow that rapid and radical change might affect wellbeing negatively, at least in the short term, because it destabilises identities, meanings, trust, relationships, etc.? The examples that we discussed in Chap. 5 suggest that this is indeed a risk, at least in the short term (e.g. a rapid collapse of institutions as witnessed in the transformation of Eastern Europe and the previous states of the Russian Federation in the early 1990s brought about considerable reductions of wellbeing), but that it also depends on societal contexts and the governance of transition.

This also links to a point we made in Chap. 7 which discusses which principles, institutions and "eco-social" policies could support wellbeing in the context of postgrowth. Here we argued that a range of policy proposals which would fit well with principles of postgrowth economies already exist in current societal niches. If integrated into a cohesive transition strategy, these may well have the potential of becoming "real utopias" (Wright 2013) to overcome contemporary capitalism. Socio-economic crises which may in the future occur in the context of "secular stagnation" might provide "windows of opportunity" to mobilise and scale up these policies.

Future research in this field will need to identify, through systematic comparative research, which kinds of contexts can support wellbeing outcomes—the fulfilment of basic human needs—in times of economic contraction and rapid change. Work has already commenced (Steinberger and Roberts 2010; O'Neill 2015; Koch et al. 2017) and is currently under way (e.g. Steinberger's "Living Well within Limits" Leverhulme Leadership project) to operationalise and identify levels of material requirements for human needs fulfilment within planetary boundaries.

Many other questions remain open and will require more detailed consideration in future research. One concerns the question of possible carriers of transition: which groups in society could be potential benefactors and advocates of postgrowth and associated alternative visions of wellbeing and policies? It seems plausible that in rich countries, people who experience precarious labour market conditions (e.g. in the growing "gig" economy, those on zero hours contracts) and who have little prospect of fitting into mainstream patterns of lifecycle achievements—including career progression, home and car ownership, etc.—could be receptive to alternative visions of life aspirations and socio-economic models beyond growth. More generally, postgrowth societies are likely to improve work–life balance and might, through the redistribution of work and other resources, weaken the male-breadwinner model and generally promote gender equality. Of course, schemes such as global cap and share of emission allowances, and the redistribution of incomes through minimum and maximum incomes would bring considerable benefits to people in developing countries and those living in poverty in the developed world. Across the world, there is already considerable support for various types of postgrowth and postgrowth-compatible movements and initiatives, demonstrated by the considerable size of audiences that postgrowth conferences have attracted over the past few years, and the various initiatives that support sustainable/slow living, voluntary simplicity, low-carbon communities, cooperatives and social enterprises, local production and exchange networks, the post-development movement, to name but a few. As argued earlier, these actors, movements and the practices and institutions they establish, can be seen as "real utopias" (Wright 2013)—societal niches which have the potential for being scaled up, especially once currently dominant practices and institutions no longer function well in the context of "secular stagnation".

References

Deaton, A. 2008. Income, Health, and Well-Being Around the World: Evidence from the Gallup World Poll. *Journal of Economic Perspectives* 22 (2): 53–72.

Doyal, L., and I. Gough. 1991. *A Theory of Human Need*. Basingstoke: Palgrave Macmillan.

Easterlin, R.A., L.A. McVey, M. Switek, O. Sawangfa, and J.S. Zweig. 2010. The Happiness-Income Paradox Revisited. *Proceedings of the National Academy of Sciences of the United States of America* 107 (52): 22463–22468.

Jackson, T., and P.A. Victor. 2016. Does Slow Growth Lead to Rising Inequality? Some theoretical reflections and numerical simulations. *Ecological Economics* 121: 206–219.

Koch, M., H. Buch-Hansen, and M. Fritz. 2017. Shifting Priorities in Degrowth Research: An Argument for the Centrality of Human Needs. *Ecological Economics* 138: 74–81.

O'Neill, D. 2015. The Proximity of Nations to a Socially Sustainable Steady-State Economy. *Journal of Cleaner Production* 108: 1213–1231.

Offe, C. 1984. *Contradictions of the Welfare State*, ed. John Keane. Cambridge, MA: MIT Press.

Steinberger, J.K., and J.T. Roberts. 2010. From Constraint to Sufficiency: The Decoupling of Energy and Carbon from Human Needs, 1975–2005. *Ecological Economics* 70 (2): 425–433.

Tversky, A., and D. Kahneman. 1991. Loss Aversion in Riskless Choice: A Reference-Dependent Model. *The Quarterly Journal of Economics* 106 (4): 1039–1061.

Wilkinson, R.G., and K.E. Pickett. 2009. *The Spirit Level. Why More Equal Societies Almost Always Do Better*. London: Allen Lane.

Wright, E.O. 2013. Transforming Capitalism Through Real Utopias. *American Sociological Review* 78 (1): 1–25.

Further Reading

Clark, B., and R. York. 2005. Carbon Metabolism: Global Capitalism, Climate Change, and the Biospheric Rift. *Theory and Society* 34 (4): 391–428.

Gasper, D. 1996. Needs and Basic Needs. A Clarification of Meanings, Levels and Different Streams of Work. Working Paper Series No. 210. The Hague: Institute of Social Studies.

Harvey, D. 2013. *Seventeen Contradictions and the End of Capitalism*. London: Profile Books.

Mont, O., and M. Koch. 2016. Conclusion: Looking Back, Looking Forward: Results and Future Research Directions. In *Sustainability and the Political Economy of Welfare*, ed. M. Koch and O. Mont, 201–212. London: Routledge.

Neumayer, E. 2013. *Weak versus Strong Sustainability: Exploring the Limits of Two Opposing Paradigms*. Cheltenham: Edward Elgar.

Parsons, T. 1968. *Structure of Social Action, a Study in Social Theory*. New York: Free Press.

Schmelzer, M. 2015. Degrowth, Klimagerechtigkeit, Subsistenz – eine Einführung in die Begriffe und Ansätze der Postwachstumsbewegung. In *Atlas der Globalisierung. Weniger wird mehr*, ed. Le Monde Diplomatique, 116–121. Berlin: Le Monde Diplomatique.

Tawney, R.H. 1938. *Religion and the Rise of Capitalism*. London: Penguin.

Victor, P.A. 2013. Growth. In *Degrowth: A Vocabulary for a New Era*, ed. G. D'Alisa, F. Demaria, and G. Kallis. London: Routledge.

M. Büchs and M. Koch, *Postgrowth and Wellbeing*,
DOI 10.1007/978-3-319-59903-8

Index

M. Büchs and M. Koch, *Postgrowth and Wellbeing*,
DOI 10.1007/978-3-319-59903-8

Printed by Printforce, the Netherlands